INTRODUCTORY
MASS
SPECTROMETRY
Second Edition

INTRODUCTORY
MASS
SPECTROMETRY
Second Edition

Stephen Shrader

CRC Press
Taylor & Francis Group
Boca Raton London New York

CRC Press is an imprint of the
Taylor & Francis Group, an **informa** business

CRC Press
Taylor & Francis Group
6000 Broken Sound Parkway NW, Suite 300
Boca Raton, FL 33487-2742

© 2014 by Taylor & Francis Group, LLC
CRC Press is an imprint of Taylor & Francis Group, an Informa business

No claim to original U.S. Government works

Printed on acid-free paper
Version Date: 20130919

International Standard Book Number-13: 978-1-4665-9584-2 (Paperback)

Library of Congress Cataloging-in-Publication Data

Shrader, Stephen R., 1942-
   Introductory mass spectrometry / Stephen Shrader. -- Second edition.
      pages cm
   "A CRC title."
   Includes bibliographical references and index.
   ISBN 978-1-4665-9584-2 (pbk. : alk. paper)
   1. Mass spectrometry. I. Title.

QC454.S515 2014
543'.65--dc23

2013036754

Visit the Taylor & Francis Web site at
http://www.taylorandfrancis.com

and the CRC Press Web site at
http://www.crcpress.com

To Marianne

Contents

List of Illustrations

List of Tables

Preface

Mass spectrometry has clearly become an integral part of the organic chemistry laboratory. Applications in both qualitative and quantitative terms have been well documented in scientific journals and books aimed at the advanced analytical or organic research chemist. This book permits the introduction of mass spectrometry and its applications at the undergraduate level. The organic chemist who wishes to add mass spectrometry to a repertoire of useful analytical tools should also find this book helpful.

Following a brief introduction to the principles and instrumentation, you will find a gradual development of the "chemistry" of mass spectrometry; the processes that occur in the mass spectrometer following ionization to give the mass spectrum are presented in terms that are familiar to the organic chemist. In the study of these chapters, it will prove useful to relate the mass spectral reactions with normal solution organic chemistry. High-resolution mass spectrometry is treated in Chapter 3.

The main intent of this book is to provide a basis for the chemist to interpret mass spectra for the solution of particular structural problems. Wherever applicable, exercises and problems are used to help clarify the discussions while providing a valuable opportunity for practicing what is learned. The examples represent a cross section of organic chemistry.

Two people played important roles in getting this second edition project off the ground. Marianne, my life and business partner for 40 years, has continually provided the inspiration to set higher goals and the drive to reach them.

My son, Philip, who has worked with us for the past 20 years, provided much of the early labor to scan the original document and ready it for editing. He then provided proofreading and suggestions leading to the final manuscript. A huge thank you to both.

Stephen Shrader
Detroit, Michigan

About the Author

Stephen Shrader, founder of Shrader Analytical and Consulting Laboratories, was among the early third-generation mass spectroscopists, starting in 1963. He is an internationally recognized expert in the use of mass spectrometry for structural elucidation and resolution of chemical problems in the petroleum, pharmaceutical, environmental and polymer fields. His work has included periods in academia, the pharmaceutical industry, and a long stint as a consultant to the overall chemical industry. Throughout his forty-plus year career, he has shared his experiences and knowledge of mass spectrometry via numerous seminars and classes. He has also been instrumental in developing a set of software tools called TSS Unity for displaying, printing, and evaluating mass spectral data for the purpose of solving chemical problems.

Abbreviations

APCI	atmospheric pressure chemical ionization
Da	dalton, the unit of mass used in mass spectrometry
DART	direct analysis in real time
EI	electron ionization
ESI	electrospray ionization
M	molecular weight of a chemical substance
m/z	mass-to-charge ratio; mass divided by the charge number
M	molecular mass of a chemical substance

1 Introduction

Mass spectrometry has played an integral part in the study of organic molecular structures for more than 50 years. Perhaps no other instrument offers as much information from so little sample as does the mass spectrometer. The mass spectrum produced by electron ionization presents a pattern of peaks that can often give definitive structural information about an unknown compound. The understanding and recognition of those patterns is the main focus of this book.

Softer ionization techniques are used to obtain definitive information regarding the molecular weight of a substance. Mass spectra obtained with these techniques are also discussed.

In the broadest sense, a spectrum is a continuum consisting of an ordered arrangement by a particular characteristic. The visible light spectrum is the range of wavelengths or frequencies of electromagnetic radiation visible to the human eye, extending from 400 (violet) to 650 nm (red). A political spectrum can be said to be a range of political ideas extending from a conservative right wing favoring individualism to a liberal left wing favoring community. A mass spectrum represents a range of molecular and fragment particles differentiated according to their mass.

Mass spectrometry is the measurement of the molecular masses and fragment masses formed from distinct chemical species.

WHAT IS MASS?

The unit of mass used in mass spectrometry is the unified atomic mass unit (symbol: u) or dalton (symbol: Da). The unit dalton is used throughout this presentation. One dalton is exactly equal to 1/12 the mass of the carbon-12 (^{12}C) isotope.

Every element consists of at least one isotope, each with its specific mass. For example, carbon has two naturally occurring isotopes:

- ^{12}C with a mass of 12.000000 Da
- ^{13}C with a mass of 13.003354 Da

Oxygen has three isotopes:

- ^{16}O with a mass of 15.994914 Da
- ^{17}O with a mass of 16.999132 Da
- ^{18}O with a mass of 17.999161 Da

Table 1.1 shows a more complete list of the isotopic masses for elements commonly found in organic chemicals.

The molecular mass (abbreviated M) of a substance is the mass of 1 molecule of that substance relative to 1 Da. This is equivalent to the older term *molecular weight*, which was abbreviated MW. The molecular mass is generally expressed as a sum of

TABLE 1.1
Exact Masses of Nuclides

Element	Atomic Weight	Nuclide	Isotopic Mass	Abundance (%)
Hydrogen	1.00797	H	1.00783	99.985
		D	2.01410	0.015
Carbon	12.01115	C	12.00000	98.90
			(standard)	
		^{13}C	13.00336	1.10
Nitrogen	14.0067	N	14.0031	99.634
		^{15}N	15.0001	0.366
Oxygen	15.9994	O	15.9949	99.762
		^{17}O	16.9991	0.038
		^{18}O	17.9992	0.200
Fluorine	18.9984	F	18.9984	100
Silicon	28.086	Si	27.9769	92.23
		^{29}Si	28.9765	4.67
		^{30}Si	29.9738	3.10
Phosphorus	30.974	P	30.9738	100
Sulfur	32.064	S	31.9721	95.02
		^{33}S	32.9715	0.75
		^{34}S	33.9679	4.21
		^{36}S	35.9671	0.02
Chlorine	35.453	Cl	34.9689	75.77
		^{37}Cl	36.9659	24.23
Bromine	79.909	Br	78.9183	50.69
		^{81}Br	80.9163	49.31
Iodine	126.904	I	126.9045	100

the masses of the most common isotope of the elements present in the molecule. For example, M for the molecule H_2O (water) is calculated in Equation 1.1:

$$M = 2*1.0078 + 15.9949 = 18.0105. \tag{1.1}$$

EXERCISE 1.1

Calculate M for the following molecules:

a. Carbon dioxide (CO_2)
b. Benzene (C_6H_6)
c. Dimethyl sulfoxide (C_2H_6SO)
d. Glucose ($C_6H_{12}O_6$)
e. Cholesterol ($C_{27}H_{46}O$)

$$M_a = 12 + 2*15.9949 = 43.9898$$

$$M_b = 6*12 + 6*1.0078 = 78.0468$$

$$M_c = 2*12 + 6*1.0078 + 31.9721 + 15.9949 = 78.0138$$

$$M_d = 6*12 + 12*1.0078 + 6*15.9949 = 180.2610$$

$$M_e = 27*12 + 46*1.0078 + 15.9949 = 386.3537$$

Note that benzene and dimethyl sulfoxide both have nominal molecular masses of 78, but the exact masses are different.

Molecular mass differs from more common measurements of the mass of chemicals, such as molar mass. Molecular mass takes into account the isotopic composition of a molecule rather than the average isotopic distribution of many molecules. As a result, molecular mass is a more precise number than molar mass.

The molar masses for the compounds in Exercise 1.1 are given in Table 1.2.

FRAGMENTATION

It was discovered quite early that polyatomic cations of even simple molecules formed in the mass spectrometer break into smaller positive ions and neutral fragments of various masses. For example, the mass spectrum of carbon dioxide (see Figure 1.1 and Table 1.3) exhibits ions with masses of 44, 28, 12, and 16 Da.

TABLE 1.2
Molar Mass and Molecular Mass

Compound	Formula	Molar Mass	Molecular Mass
Carbon dioxide	CO_2	44.01	43.9898
Benzene	C_6H_6	78.11	78.0468
Dimethyl sulfoxide	C_2H_6SO	78.13	78.0138
Glucose	$C_6H_{12}O_6$	180.16	180.2610
Cholesterol	$C_{27}H_{46}O$	386.65	386.3537

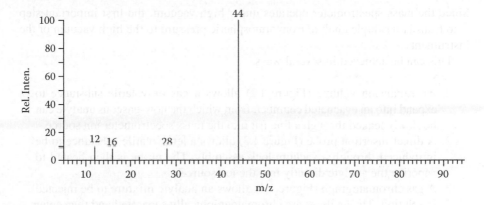

FIGURE 1.1 Mass spectrum of carbon dioxide.

TABLE 1.3

Masses of CO_2 and its Fragments

Ion	Nominal Mass
CO_2^+	44
CO^+	28
C^+	12
O^+	16

It is not difficult to recognize that these ions correspond to the carbon dioxide molecule, carbon monoxide molecule, carbon atom, and oxygen atom, respectively, as depicted by Equation 1.2.

$$CO_2 \xrightarrow{\text{electron energy}} CO_2^+ + CO^+ + C^+ + O^+ \qquad (1.2)$$

Not until the late 1950s, however, were significant attempts made to correlate these fragment ions with molecular structures. And, in the ensuing years, mass spectra of literally millions of organic compounds containing various functional groups have been examined and successfully related to molecular structure.

THE MASS SPECTROMETER

The basic functions of a mass spectrometer are

1. To introduce a chemical analyte into the high vacuum of the instrument;
2. To produce ions from that analyte;
3. To separate the ions according to their masses; and
4. To measure the abundances of those ions.

SAMPLE INTRODUCTION

Since the mass spectrometer operates under high vacuum, the first important step is to transfer a sample analyte from atmospheric pressure to the high vacuum of the instrument.

This can be achieved in several ways.

1. An expansion volume (Figure 1.2) allows a gas or volatile substance to expand into an evacuated chamber, from which the now-gaseous analyte can be slowly leaked through a fine frit into the mass spectrometer ion source.
2. A direct insertion probe (Figure 1.3) allows a less-volatile substance to be introduced through a vacuum-lock assembly. The probe is then heated to vaporize the analyte directly into the ion source.
3. A gas chromatograph (Figure 1.4) allows an analyte mixture to be injected in solution. The analytes are chromatographically separated and then enter the mass spectrometer individually as pure substances.

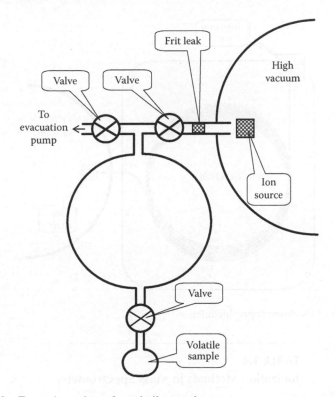

FIGURE 1.2   Expansion volume for volatile samples.

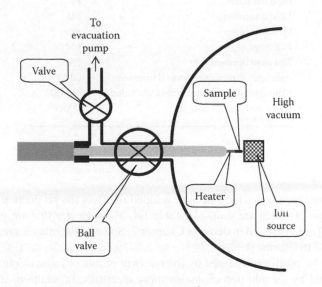

FIGURE 1.3   Direct insertion probe in vacuum lock.

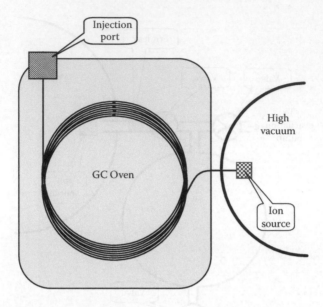

FIGURE 1.4   Gas chromatographic inlet.

TABLE 1.4

Ionization Methods in Mass Spectrometry

Ionization Method	Abbreviation
Electron ionization	EI
Chemical ionization	CI
Field ionization	FI
Field desorption	FD
Photoionization	PI
Electrospray	ESI
Fast atom bombardment	FAB
Atmospheric pressure chemical ionization	APCI
Matrix-assisted laser desorption ionization	MALDI
Direct analysis in real time	DART

IONIZATION

Many different techniques have been developed to effect the production of ions, the most common of which are shown in Table 1.4. Mass spectra that are generated by electron (EI) are discussed in detail in Chapter 2. Some of the other ionization modes are discussed in Chapter 4.

Ions may be positively charged by the removal of one or more electrons or negatively charged by the addition of one or more electrons. In addition, smaller mass ions are formed by various fragmentation processes discussed in the next chapter.

These ions are then accelerated by electrostatic fields into a mass filter to be separated and detected. The entire process takes place under high vacuum to minimize collisions between ions.

MASS SEPARATION

The ions formed in the mass spectrometer are separated according to their masses or, more correctly, according to their mass-to-charge ratios (*m/z*), as shown in Equation 1.3:

$$\frac{m}{z} = \frac{(Mass\ of\ ion)}{(\#\ of\ missing\ or\ additional\ electrons)} \tag{1.3}$$

The techniques to effect the separation of ions according to their mass-to-charge ratio are also varied. Among them are the following:

- Time of flight (Figure 1.5), by which masses are separated according to the time they take to traverse a given distance in a vacuum. Ions are pulsed into an acceleration region, where they gain momentum. They then enter a field free drift region, where ions of lower mass travel faster than heavier ions, thus providing mass separation. A series of focusing plates, referred to as the reflectron, bends the path of the ion beam to obtain a longer flight path.
- Quadrupole, by which only a specific mass passes through a set of parallel rods with oscillating electric fields superimposed on a DC (direct current) field. Varying the frequency of oscillation or the DC voltage changes the mass that passes through to the detector (Figure 1.6).

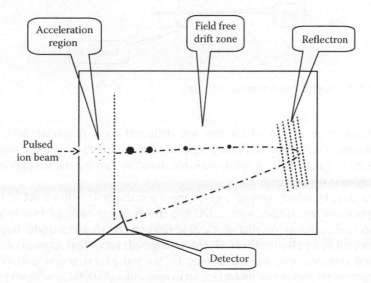

FIGURE 1.5 Time-of-flight mass analyzer.

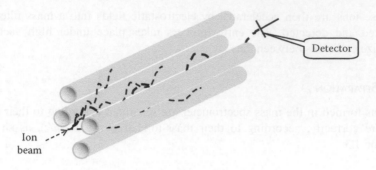

FIGURE 1.6   Quadrupole mass analyzer.

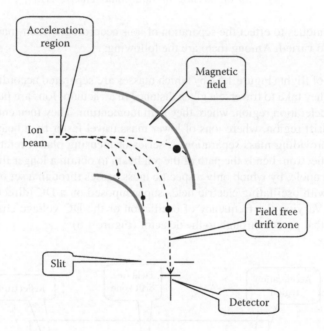

FIGURE 1.7   Magnetic sector mass analyzer.

- Magnetic sector, in which ions are deflected by a magnetic field. The amount of deflection is directly dependent on the ion mass and the strength of the magnetic field. A spectrum is measured by varying the strength of the magnetic field (Figure 1.7).
- Ion trap, by which ions of a given mass are trapped within a set of three hyperbolic electrodes with a DC electric field on each of two end-cap electrodes and an oscillating electric field on the ring electrode. Ions are injected in a pulsed manner, stored (or trapped) for a short amount of time, then extracted one mass at a time by the pulsed extraction grid. These instruments can detect molecules up to molecular 70,000 Da with very high sensitivity (Figure 1.8).

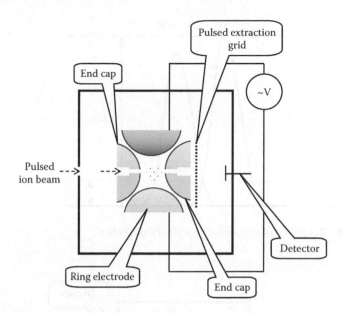

FIGURE 1.8   Ion trap mass analyzer.

The resultant mass spectra are nonetheless very similar to one another. The primary differences are in the ultimate mass resolution or the difference in mass between two ions that can just be separated from one another.

RESOLUTION

The function of the mass analyzer is to separate an ion of mass M from one of mass $M + \Delta M$. The resolution R of an instrument is defined by Equation 1.4:

$$Resolution\,(R) = \frac{M}{\Delta M} \tag{1.4}$$

ΔM is the difference in mass between two ions separated at mass M with less than 10 percent peak overlap, as shown by Figure 1.9.

Rather arbitrarily, low-resolution instruments can be defined as those that separate unit masses up to m/z 2,000 (the resolution is 2,000/1 = 2,000). Unit mass (or low-resolution) spectra are obtained from these instruments.

An instrument is generally considered high resolution if it can separate two ions differing in mass by at least 1 part in 5,000 to 15,000 (R = 5,000- 15,000). An instrument with 10,000 resolution can separate an ion of mass 200.00 from one of mass 200.02 (R = 200/0.02 = 10,000). This important class of mass spectrometers allows the measurement of the exact masses of ions, which in turn leads to elemental compositions of ions. Chapter 3 deals with high-resolution mass spectrometry and accurate mass measurement in detail.

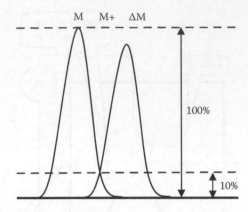

FIGURE 1.9 Resolution of two ions with a 10 percent valley.

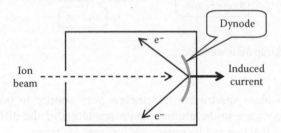

FIGURE 1.10 Faraday cup detector.

Ion Detection

To accurately measure the abundances of the ions formed and separated in the mass spectrometer, a reliable detector is an important part. The following detectors are commonly used in commercial instrumentation.

1. The Faraday cup (Figure 1.10) is the simplest, with a secondary electron current generated by ions striking a dynode surface coated with an emitting material such as CsSb, GaP, or BeO. The resulting impact emits electrons and produces an induced current that can be amplified. This detector is not particularly sensitive but is very stable and is ideally suited for isotope analysis.
2. The electron multiplier (Figure 1.11) is a series of dynodes that serve to multiply the output current manyfold. An ion striking the first dynode causes a few electrons to be emitted, which in turn strike the next dynode, causing more electrons to be emitted. The electron current at the final dynode will typically be multiplied by 1 million or more.
3. The photomultiplier detector (Figure 1.12) is similar to the electron multiplier, except photons emitted from a phosphorescent screen provide the multiplying effect.

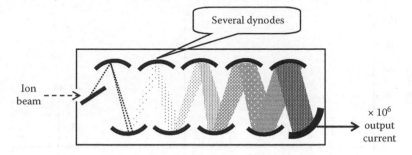

FIGURE 1.11   Electron multiplier detector.

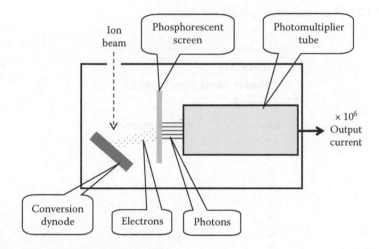

FIGURE 1.12   Photomultiplier detector.

DATA PRESENTATION

For clarity of presentation, mass spectra may be in the form of a bar graph or a mass and abundance table.

Bar Graph

Figure 1.13 demonstrates the graphical method of displaying the same mass spectrum. The x-axis has the unit of mass-to-charge ratio (*m/z*); the y-axis is ion intensity, relative to the intensity of the most abundant ion.

The bar graph is preferred when the general appearance of a spectrum is an important factor in interpretation.

Tabular

A table offers the advantage of accurate mass and ion abundance presentation. Table 1.5 demonstrates the tabular method of displaying a mass spectrum.

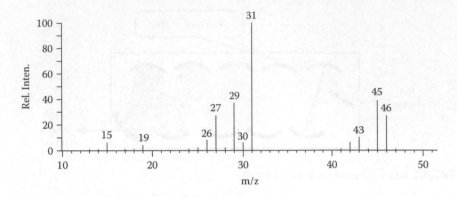

FIGURE 1.13   Bar graph mass spectrum of ethanol.

TABLE 1.5
Tabular Mass Spectrum of Ethanol

m/z	Relative Intensity[a]
15	6.7
19	2.8
26	9.9
27	22.5
28	3.5
29	29.9
30	8.2
31	100
42	4.8
43	11.5
45	51.5
46	21.7

[a] Based on the most intense peak (m/z 31) arbitrarily set at 100.

Chapter 2 of this book describes the mass spectra produced by EI ionization and how the spectral peak pattern relates to molecular structure. Chapter 3 describes in detail the use of high-resolution and accurate mass measurement to determine elemental compositions of ions as an aid to identifying unknown substances. Chapter 4 provides an introduction to some of the more modern techniques used to extend the usefulness of mass spectrometry into high molecular weight and more polar substances.

2 Electron Ionization

ION FORMATION

The reaction scheme shown in Equation 2.1 depicts how electron (EI) ionization employs a high-energy electron beam to excite molecules M in the vapor state to high electronic and vibrational energy levels M^*. When enough energy is supplied to a molecule, an electron is ejected to produce a positively charged species M^+. This is termed the molecular ion.

$$M \xrightarrow{\text{electron energy}} M^* \xrightarrow{\text{more energy}} M^+ \xrightarrow{\text{more energy}} A^+ + B^+ + C^+ \qquad (2.1)$$

When the electron energy is raised above the minimum required to ionize the molecule, the molecular ions will be more apt to fragment. Electron energies on the order of 50 to 70 electron volts (eV) are generally used in routine mass spectrometric tests.

The mass spectrum of ethane (Figure 2.1) shows that nearly every possible fragment is formed. The positive ions $CH^+(13)$, $CH_2^+(14)$, $CH_3^+(15)$, $C_2H^+(25)$, $C_2H_2^+(26)$, $C_2H_3^+(27)$, $C_2H_4^+(28)$, $C_2H_5^+(29)$, and $C_2H_6^+(30)$ are present in the EI mass spectrum of ethane. Only the most abundant ions, however, are generally considered important for interpretive purposes.

EXERCISE 2.1
Make a list of those ions formed from ethane that have abundances greater than 10 percent of that of the most abundant ion.

- The ion that appears at m/z 30 corresponds to the molecular ion $C_2H_6^+$.
- Fragment ions formed by the loss of 1, 2, 3, and 4 hydrogen atoms appear at m/z 29, 28, 27, and 26, respectively.
- Peaks arising from lower-mass ions are of low intensity.
- 30, 29, 28, 27, and 26.

TYPES OF IONS

MOLECULAR ION

A molecular ion is the ion formed by the simple loss of one electron from a molecule. The most fundamental structural information provided by a mass spectrum is the molecular weight of a compound, as defined by the mass of the molecular ion.

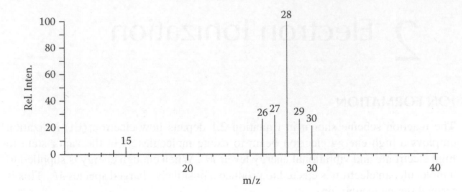

FIGURE 2.1   Mass spectrum of ethane.

FRAGMENT ION

Fragment ions result from cleavage of one or more bonds in the molecular ion. They represent building blocks that, if put together properly with the neutral fragments, can describe the complete structure of the original molecule.

MULTIPLY-CHARGED IONS

Both molecular and fragment ions may have more than a single missing electron ($z = 2, 3, ...$). Doubly charged ions are fairly common, but ions with multiple charges greater than 2 are observed only rarely.

Doubly charged ions appear in the mass spectrum at half of their actual mass. More examples are shown in Table 2.1.

It is important in the interpretation of spectra to identify the various types of ions in a mass spectrum. Assignment of one peak to the molecular ion is an important first step in the interpretation of most spectra; the remainder of the peaks will correspond to fragment ions. This chapter describes how the various ion types may be recognized and used in the interpretation process.

MOLECULAR IONS

The term *molecular ion* by now requires little clarification; it simply describes an ion formed by the removal of one electron from a molecule. If all the bonds in a molecular ion are sufficiently strong, the energy used for ionization will not disrupt the atoms, and an intense molecular ion peak will be observed. If they are not, fragmentation of the weaker bonds will occur to give fragment ions. In these cases, a molecular ion peak may or may not be observed, depending on the rate at which the molecular ion decomposes. This rate in turn depends on the bond strengths within the ion, the facility with which the positive charge is delocalized (and thus stabilized), the stabilities of possible fragment ions, and the stabilities of neutral species formed in the fragmentations.

TABLE 2.1

Values of *m/z* for Multiply-charged Ions

Mass (*m*)	Change (*z*)	*m/z*
100	1	100
75	1	75
220.2	1	220.2
565.4	1	565.4
323.3	1	323.3
210.2	2	105.1
75	2	37.5
220.2	2	110.1
455.4	2	227.7
692.6	2	346.3
129	3	43
366.3	3	122.1
693.6	3	231.2
12,000	100	120

Several characteristics of molecular ions can be used to distinguish them from others:

1. A molecular ion peak must have the highest mass of any peak in the spectrum.

    Under normal operating conditions, the mean free path of the ions and molecules is great enough to preclude a significant number of bimolecular or ion-molecule collisions, and only by the transfer of atoms to a molecular ion during such a collision can a particle heavier than the molecular weight be formed.

    The converse, however, is not true: *The highest mass peak need not correspond to a molecular ion.*

    A molecular ion may not be observed if one or more bonds are too weak to withstand the harsh ionization conditions of the electron beam.

    Certain structural features may highly favor the cleavage of a specific bond so a stabilized ion may be formed. For example, cleavage of a bond adjacent to a heteroatom is a favorable process. The stable oxonium ion *A*, shown in Equation 2.2, is formed from ethylene ketals, and molecular ion peaks are seldom observed.

$$
\begin{array}{ccc}
\text{R} \diagdown \diagup \text{R}' & & \text{R} \\
\quad\times\quad & \xrightarrow[\; -\text{R}' \;]{\text{Electron energy}} & \| \\
\text{O} \quad\quad \text{O} & & \text{O} \diagdown \diagup^{+} \text{O} +
\end{array}
\qquad (2.2)
$$

As another example, aliphatic hydrocarbons contain a large number of bonds, all with nearly equal energies. The likelihood that at least one of the bonds will break is high, so molecular ion peaks from aliphatic hydrocarbons will be of low intensity if observed at all. Cyclic hydrocarbons, on the other hand, must cleave two carbon-carbon bonds before any loss of mass occurs (neglecting the relatively unimportant loss of a hydrogen atom), with the result of greatly increased molecular ion intensities.

A third situation in which a molecular ion peak may not be observed is when a fairly stable fragment ion can be formed along with a very stable neutral molecule or radical. The elimination of water from alcohols, for example, often occurs readily so that the highest-mass peak in the spectrum corresponds to an olefin fragment with a mass that is 18 mass units lower than the actual molecular weight, as depicted by the reaction scheme shown in Equation 2.3.

$$ROH \xrightarrow{\ \ electron\ energy\ \ } [M-18]^+ \tag{2.3}$$

2. The mass of a molecular ion must be even unless the molecule contains an odd number of nitrogen atoms.

The molecular weights of the compounds shown in Table 2.2 demonstrate this rule.

Nitrogen is the only commonly encountered element that has an even atomic mass yet an odd-number valency. Thus, any compound not containing nitrogen must have an even molecular weight.

Deuterium also falls into this odd class of elements, so in work with isotopically labeled materials, the stated rule is not valid.

From other information concerning the molecular structure, such as isotope peak calculations, exact mass measurements, or spectral or chemical evidence, the presence or absence of nitrogen may be known. A compound known to be a steroid, for example, must have an even molecular weight. On the other hand, a substance that becomes water soluble under acidic conditions must contain at least one basic nitrogen function; in this case, the molecular weight may be odd or even, depending on the number of nitrogen atoms actually present.

3. The peak at next lower mass must not correspond to the loss of an impossible or improbable combination of atoms.

For example, a peak 5 mass units lower than a proposed molecular ion could only be formed by the loss of 5 hydrogen atoms, but a molecule of hydrogen (H_2, 2 Da) is the maximum number of hydrogen atoms lost in a single fragmentation step. Thus, a peak 5 Da lower than the molecular ion is not a reasonable fragmentation ion.

The smallest fragment that may be lost is the hydrogen atom (H, 1 Da). The next smallest fragment is a hydrogen molecule (H_2, 2 Da). The next smallest fragment that can be lost is the methyl radical (CH_3, 15 Da). Therefore, two ions differing by 3 to 14 Da must be either

TABLE 2.2
Examples of Structures, Formulas, and Molecular Masses

Compound Name	Structure	Formula	M
Cholesterol		$C_{27}H_{46}O$	386
Nitrobenzene		$C_6H_5NO_2$	123
Thiophene		C_4H_4S	84
Tetramethylsilane		$C_4H_{12}Si$	88
Aminopyridine		$C_4H_6N_2$	94
Fluorobenzene		C_6H_5F	96
Triphenylphosphine		$C_{18}H_{15}P$	262
Guanidine		CH_5N_3	59

a. Fragments from a molecular ion of higher mass, or
b. From a mixture of two different compounds.

Other mass differences may be judged to be impossible based on information already derived from the spectrum.

A compound containing only carbon, hydrogen, nitrogen, and oxygen, for example, cannot lose a fragment with a mass between 19 and 24 Da.

By similar reasoning, if the isotope peak patterns (discussed in the next section) demonstrate the absence of chlorine, two peaks from 35 to 38 Da apart cannot correspond to molecular and fragment ions of the same compound.

A common occurrence, especially with compounds isolated from natural sources, is to see two peaks in the molecular ion region separated by only 14 mass units. These are not from a single compound but are molecular ions of homologs that differ in structure and molecular weight by a methylene (CH_2, 14 Da) group. The similarity of such compounds often makes it extremely difficult to achieve chemical or physical separation or even to detect such a mixture by methods other than mass spectrometry.

4. No fragment ion may contain a larger number of atoms of any particular element than the molecular ion.

Such an observation could only be made if collisions within the ion source transferred atoms from one ion (or molecule) to another. We have already excluded this possibility by virtue of the low pressures in the ion source. The application of this rule takes on added significance in discussions of high-resolution mass spectrometry in Chapter 3.

ISOTOPE PEAKS

Table 2.3 shows the elements commonly found in organic compounds and the relative abundances of their isotopes. Of these elements, only fluorine, phosphorus, and iodine are monoisotopic. Most elements are mixtures of two or more stable isotopes, differing in mass by 1 or 2 Da. Most elements include one major isotope (greater than 90 percent relative abundance), but chlorine and bromine have two rather abundant isotopes separated by 2 Da.

By virtue of their different masses, the isotopes result in separate peaks in the mass spectrometer, with the intensities of these peaks directly proportional to the abundances of the isotopes. Isotope peak intensities therefore provide important information regarding the elemental compositions of ions.

Note in Table 2.3 that the elements can be divided into three groups:

1. Monoisotopic elements: fluorine, phosphorus, and iodine
2. Elements occurring in large numbers in a molecule but with low abundances of heavy isotopes: hydrogen, carbon, oxygen, and nitrogen
3. Elements occurring in small numbers but with high abundances of heavy isotopes: silicon, sulfur, chlorine, and bromine

TABLE 2.3
Isotopic Distribution of Common Elements

Element	Symbol	Nominal Mass	Exact Mass	Relative Abundance
Hydrogen	H	1	1.007825	99.99
	D or ^{2}H	2	2.014102	0.01
Carbon	C	12	12.000000	98.9
	^{13}C	13	13.003354	1.1
Nitrogen	N	14	14.003074	99.6
	^{15}N	15	15.000108	0.4
Oxygen	O	16	15.994914	99.76
	^{17}O	17	16.999132	0.04
	^{18}O	18	17.999161	0.2
Fluorine	F	19	18.998403	100
Silicon	Si	28	27.976927	92.2
	^{29}Si	29	28.976495	4.7
	^{30}Si	30	29.973770	3.1
Phosphorus	P	31	30.973762	100
Sulfur	S	32	31.972071	95.0
	^{33}S	33	32.971459	0.75
	^{34}S	34	33.967867	4.25
Chlorine	Cl	35	34.968852	75.76
	^{37}Cl	37	36.965902	24.24
Bromine	Br	79	78.928337	50.69
	^{81}Br	81	80.916291	49.31
Iodine	I	127	126.904473	100

As a first approximation, the contributions to isotope peak intensities by various elements are additive. Each group of elements can be treated separately and the results summed.

GROUP 1

There is no contribution to isotope peak intensities by these elements. Therefore, weak isotope peak intensities for relatively high-mass ions are evidence for the presence of monoisotopic elements.

GROUP 2

In simple molecules containing only carbon, hydrogen, nitrogen, oxygen, and the monoisotopic elements, elemental compositions of ions may be calculated using isotope peak intensities. As a first approximation, one atom in an ion contributes an amount to the intensity of the isotope peaks that is equal to the relative abundances of the isotopes of that atom. When more than one atom is present, the intensity contribution is multiplied by the number of each atom present.

This approximation is restated by Equations 2.4 and 2.5. If P is the intensity of the ion with no heavy isotope, then $(P + 1)$ is the intensity of the peak 1 mass higher, and $(P + 2)$ is the intensity of the peak 2 masses higher.

An additional term is added to the equation for $(P + 2)$ to account for contributions of two ^{13}C atoms in a single ion; this term becomes important when there are a large number of carbon atoms in an ion.

$$(P + 1) = 1.1 * (\# C\ atoms) + 0.37 * (\# N\ atoms) \tag{2.4}$$

$$(P + 2) = 1.1 * \frac{(\# C\ atoms)^2}{200} + 0.37 * (\# O\ atoms) \tag{2.5}$$

GROUP 3

Examine the isotopic mixture of chlorine. The molecular ion of ethyl chloride (see Figure 2.2) is revealed as two peaks separated by 2 Da and with an approximate intensity ratio corresponding to the relative abundances of the two isotopes (3:1). The lower mass peak of this doublet at 64 Da is referred to as P and corresponds to $C_2H_5Cl^{35}$, and the higher mass peak at 66 Da is $(P + 2)$ and corresponds to $C_2H_5Cl^{37}$. A pair of peaks separated by two masses and with an intensity ratio of 3:1 is therefore characteristic of one chlorine atom in a molecule. Likewise, a molecular ion containing bromine appears as two peaks separated by two mass units and with nearly equal intensities (Figure 2.3 is the mass spectrum of ethyl bromide).

EXERCISE 2.2

Identify ions other than the molecular ions that contain halogen atoms in the mass spectra shown in Figures 2.2 and 2.3.

X^+ ions (corresponding to the ionized halogen atoms) appear at m/z 35(37) and 79(81) in the two spectra; HX^+ ions are at m/z 36(38) and 80(82). There are similar

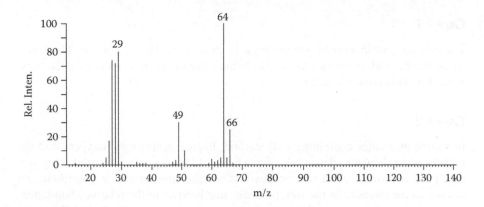

FIGURE 2.2   Mass spectrum of ethyl chloride.

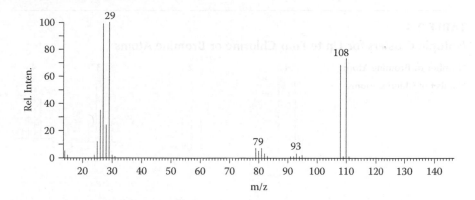

FIGURE 2.3 Mass spectrum of ethyl bromide.

doublets 14 mass units above the X^+ ions in both spectra corresponding to CH_2Cl^+ (m/z 49, 51) and CH_2Br^+ (m/z 93, 95) ions.

When an ion contains more than one atom from group 3, the isotopic distribution can be determined from expansion of the binomial expression in Equation 2.6:

$$(A + B)\wedge n \tag{2.6}$$

where

A = relative abundance of the light isotope
B = relative abundance of the heavy isotope
n = number of atoms present in a molecule

For example, for two chlorine atoms ($A \sim 3$, $B \sim 1$, $n = 2$), expansion of expression 2.3 gives

$$(A + B)^n = A^2 + 2*AB + B^2 = 9 + 6 + 1$$

The isotope peak intensity ratio will therefore be 9:6:1.

Similarly, for three bromine atoms ($A \sim 1$, $B \sim 1$, $n = 3$),

$$(A + B)^n = A^3 + 3*A^2*B + 3*A*B^2 + B^3 = 1 + 3 + 3 + 1$$

The isotope peak intensity ratio will therefore be 1:3:3:1.

Table 2.4 shows the results of these calculations for up to four chlorine or bromine atoms.

Silicon and sulfur also contribute somewhat to a ($P + 2$) isotope peak. As a first approximation, one atom in an ion contributes an amount to the intensity of the

TABLE 2.4
Isotopic Clusters for Up to Four Chlorine or Bromine Atoms

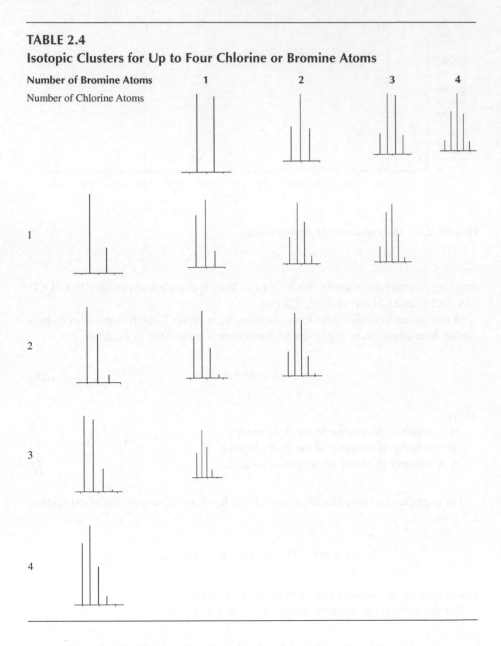

| Number of Bromine Atoms | 1 | 2 | 3 | 4 |

Number of Chlorine Atoms

1

2

3

4

isotope peaks that is equal to the relative abundances of the isotopes of that atom. When more than one atom is present, the intensity contribution is multiplied by the number of each atom present.

When evaluating isotope peak intensities, the first step is to compare the intensity of the $(P + 2)$ peak to the relative abundances of the heavy isotopes of bromine, chlorine, sulfur, and silicon. If there is a fairly close fit (with one or more atoms assumed present), the $(P + 1)$ intensity is corrected by subtracting contributions to it by Si^{29} or S^{33}. The number of carbon atoms can now be determined in the usual manner by dividing the $(P + 1)$ intensity by 1.1, the relative abundance of carbon 13.

For complex compositions, isotope peak calculations are best performed using any of the mass spectrometry tools available in most software or on the Internet.

EXERCISE 2.3

What isotope peak intensities are expected for the molecular ions of aniline (C_6H_7N), acetophenone (C_8H_8O), ethyl iodide (C_2H_5I), and ethylamine (C_2H_7N)?

Aniline

$$P = 100$$

$$(P + 1) = 1.1*6 = 0.37 = 7.0$$

$$(P+2) = \frac{6.6^2}{200} = 0.21$$

Acetophenone

$$P = 100$$

$$(P + 1) = 1.1*8 = 8.8$$

$$(P + 2) = 8.82/200 + 0.2 = 0.58$$

Ethyl iodide

$$P = 100$$

$$(P + 1) = 1.1*2 = 2.2$$

Ethylamine

$$P = 100$$

$$(P + 1) = 1.1*2 + 0.37 = 2.6$$

PROBLEM 2.1

What are the elemental compositions of the compounds whose spectra are in Figures 2.4 through 2.9? Identify the substances. Molecular ions are observed in all cases.

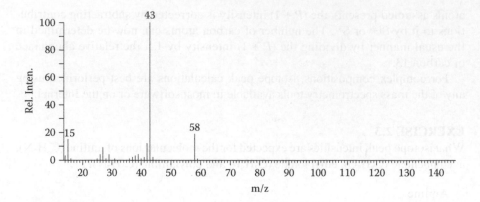

FIGURE 2.4 Problem 2.1a.

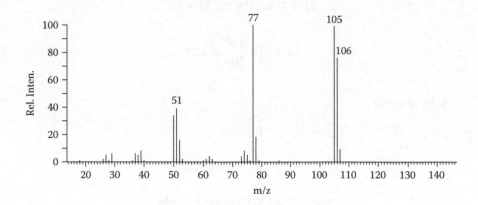

FIGURE 2.5 Problem 2.1b.

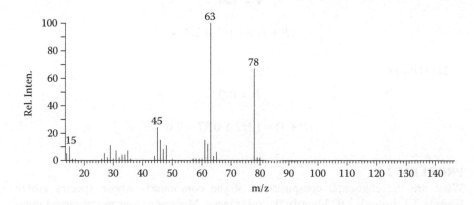

FIGURE 2.6 Problem 2.1c.

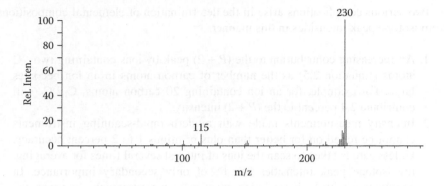

FIGURE 2.7 Problem 2.1d.

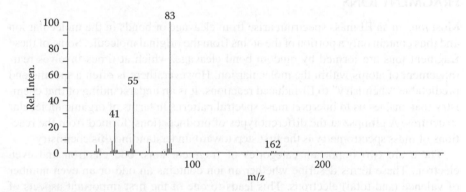

FIGURE 2.8 Problem 2.1e.

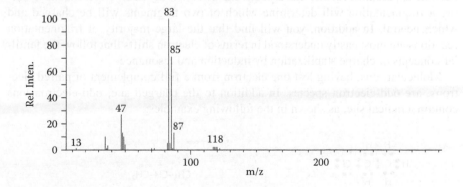

FIGURE 2.9 Problem 2.1f.

Two serious complications arise in the determination of elemental compositions from isotope peak intensities in this manner.

1. An increasing contribution to the $(P + 2)$ peak by ions containing two ^{13}C atoms (Equation 2.5) as the number of carbon atoms in an ion becomes large. For example, for an ion containing 20 carbon atoms, $C_{18}{}^{13}C_2$ will contribute 2.4 percent to the $(P + 2)$ intensity.
2. Intensity measurements made with modern rapid-scanning instruments cannot be relied on for better than plus or minus 1 to 2 percent accuracy. Unless care is taken to scan the ions of interest several times for averaging, the isotope peak intensities can be of only secondary importance. In conjunction with high-resolution exact mass measurements, however, the approximate isotope peak intensities can be of considerable value. This will become clear as you proceed through this book.

FRAGMENT IONS

Most ions in an EI mass spectrum arise from cleavage of bonds in the molecular ion and thus contain only a portion of the atoms from the original molecule. Some of these fragment ions are formed by random bond cleavage, which at times involves rearrangement of atoms within the molecular ion. However, there is often a specific and predictable "chemistry" to EI-induced reactions. It is an understanding of that chemistry that enables us to interpret mass spectral patterns in terms of organic molecular structures. A glimpse at the different types of products (ions) formed from the reactions of mass spectrometry is the first step toward understanding this chemistry.

There are two important categories of fragment ions: odd electron and even electron. These terms describe whether an ion contains an odd or an even number of valence (and total) electrons. This leads to one of the first important aspects of mass spectral interpretation: electron counting or electron bookkeeping. Although the mechanisms drawn to account for observed fragmentations (and particularly the movement of electrons within an ion) are highly speculative, they are helpful for keeping a close check on the electrons in a molecule and subsequent ions. As should become apparent in further discussions, the direction in which electrons move during a fragmentation will determine which of two fragments will be charged and which neutral. In addition, you will find that the large majority of fragmentation reactions are most easily understood in terms of electron shifts that follow the familiar concepts of charge stabilization by induction and resonance.

Molecular ions, having lost one electron from a full complement of paired electrons, are odd-electron species. In addition to the charged site, odd-electron ions contain a radical site, as shown in the following examples:

M$^+$ Ethyl chloride

$$\overset{+}{CH_3-CH}-\overset{\bullet}{CH_2}$$

M$^+$ Propylene

M+ Butadiene   M + Benzene

It is emphasized that the electronic or, for that matter, the nuclear structures of ions cannot be unambiguously ascertained. Structures are indicated with localized charge and radical site throughout this book merely as an aid to interpreting mass spectra and fragmentation processes that occur.

Fragmentation of molecular ions (or any odd-electron ion) may occur by cleavage of bonds in two ways: heterolytic or homolytic.

Heterolytic cleavage is designated by a conventional arrow $\curvearrowright$ to signify the transfer of a pair of electrons in the direction of the charged site (see Equation 2.7).

$$CH_3CH_2 \overset{\curvearrowleft}{-} \overset{\bullet \,+}{\underset{\bullet\bullet}{Cl}} \longrightarrow CH_3CH_2^+ + \overset{\bullet\bullet}{\underset{\bullet\bullet}{Cl}} \qquad (2.7)$$

o.e.   e.e.   o.e.

Homolytic cleavage is designated by a fishhook arrow $\curvearrowright$ to signify the transfer of a single electron (Equation 2.8).

$$CH_3 \overset{\curvearrowleft}{-} CH_2 \overset{+}{C}H - \overset{\bullet}{C}H_2 \longrightarrow CH_2 \overset{\cdots}{\underset{CH}{}} + CH_2 + \overset{\bullet}{C}H_3 \qquad (2.8)$$

o.e.   e.e.   o.e.

These examples involve the loss of odd-electron neutral fragments and thus result in even-electron ions. This is always the case in simple cleavage fragmentations, a process in which only one bond is broken.

Cleavage of two bonds, either consecutively or simultaneously, results in even-electron neutral fragments and odd-electron ions (Equation 2.9).

$$\begin{array}{c} H_2C - CH_2 \\ H \overset{\curvearrowright}{\underset{\bullet\bullet}{Cl^+}} \end{array} \longrightarrow \overset{\bullet}{C}H_2 - \overset{+}{C}H_2 + HCl \qquad (2.9)$$

o.e.   e.e.   o.e.

Cleavage of three bonds again produces an even-electron ion. It will become evident in further discussions that odd-electron fragment ions may be mistakenly identified as molecular ions. For example, the M–H$_2$O ion of alcohols is an odd-electron fragment ion that corresponds to a molecular ion of the related alkene (Equation 2.10).

$$R - \underset{\overset{|}{CH}}{} - CH_2 \overset{\curvearrowright}{\underset{\bullet\,+}{OH}} \overset{-H_2O}{\longrightarrow} R - \overset{\bullet}{C}H - \overset{+}{C}H_2 \qquad (2.10)$$

M$^+$ (o.e.)   o.e.

Most ions in a mass spectrum are even-electron ions, formed by cleavage of a single bond in the molecular ion. Decomposition of these ions may proceed in two ways:

1. Loss of an even-electron (e.e.) neutral fragment or molecule to yield an even-electron ion.

$$H_3C \overbrace{—CH_2—}CH_2^+ \longrightarrow CH_3^+ + CH_2 = CH_2 \qquad (2.11)$$
$$\text{e.e.} \qquad\qquad \text{e.e.} \qquad \text{e.e.}$$

2. Loss of an odd-electron radical to give an odd-electron (o.e.) ion (Equation 2.12). This type of fragmentation is far less prevalent than the first. Note that the nature of the product ion is not necessarily dependent on the number of bonds cleaved but on whether heterolytic or homolytic cleavage occurs.

$$H_3C \overbrace{—CH_2—}CH_2^+ \longrightarrow \overset{\bullet}{C}H_2 — \overset{+}{C}H_2 + CH_3 \qquad (2.12)$$
$$\text{e.e.} \qquad\qquad \text{o.e.} \qquad \text{o.e.}$$

The processes described are succinctly summarized by Equations 2.13 to 2.16.

$$o.e.^+ \rightarrow e.e.^+ + o.e. \qquad (2.13)$$

$$o.e.^+ \rightarrow o.e.^+ + e.e. \qquad (2.14)$$

$$e.e.^+ \rightarrow e.e.^+ + e.e. \qquad (2.15)$$

$$e.e.^+ \rightarrow o.e.^+ + o.e. \qquad (2.16)$$

MULTIPLY-CHARGED IONS

The last (but certainly not least) important group of ions encountered in mass spectrometry includes those having more than a single positive charge. From Equation 1.3, the "mass" of a multiply charged ion will appear to be some fraction of its actual mass. Thus, the doubly charged molecular ion of benzene (molecular weight 78) appears at m/z 39, the same mass as the singly charged $C_3H_3^+$ ion. Although triply charged ions are seldom formed, they appear at one-third of their actual mass.

Identification of the doubly charged ions in a mass spectrum is not particularly difficult.

- Odd-mass doubly charged ions appear at a "half-mass" position. A doubly charged ion of mass 115 will appear at m/z 57.5.
- Even-mass doubly charged ions are identified by the ^{13}C isotope peak and appear at a mass 0.5 Da greater than the P ion. For example, the doubly charged ion of mass 130 will appear at m/z 65, with the ^{13}C isotope peak at m/z 65.5.

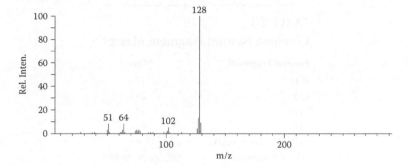

FIGURE 2.10 Mass spectrum of naphthalene.

Doubly charged ions are of greatest importance in the mass spectra of highly unsaturated (particularly aromatic) compounds. The mass spectrum of naphthalene (Figure 2.10) has groups of peaks around the following:

- m/z 128 [M]$^+$ and 64 (M^{++})
- m/z 102 [M–C$_2$H$_2$]$^+$ and 51 [M–C$_2$H$_2$]$^{++}$
- m/z 76 [M–2C$_2$H$_2$]$^+$ and 38 [M–2C$_2$H$_2$]$^{++}$

In certain other cases, the formation of doubly charged ions may also be highly favored. The most intense peak in the spectra of the complex naturally occurring bisbenzyltetrahydroisoquinoline alkaloids is due to the doubly charged ion a. The two nitrogen atoms each promote cleavage of the *alpha* bond, and since the centers are so remote from each other, the double charge is easily accommodated by the fragment (Equation 2.17). The repulsion between like charges is small at such large intramolecular distances.

(2.17)

TABLE 2.5

Common Neutral Fragment Masses

Neutral Fragment	Mass
H_2O	18
CO	28
CO_2	44
SO_2	64
ROH (R = alkyl, acyl)	32, etc.
HX (X = halogen)	20, 36, 80, or 128

NEUTRAL FRAGMENTS

For every positively charged fragment ion, at least one neutral fragment is also formed. These fragments are unaffected by the electric and magnetic fields of the mass spectrometer and therefore cannot be studied in the normal ways; their structures and properties can only be ascertained by examining the ions with which they are formed.

Fragmentation reactions are often promoted by the formation of certain stable neutral fragments or molecules. The most common of these stable species are shown in Table 2.5. They are discussed in greater detail in the next section.

BASIC MASS SPECTRAL REACTIONS

INTRODUCTION

For many years, organic mass spectrometry suffered from the misconception that molecules fall apart randomly after being bombarded by high-energy electrons. The energy imparted to the sample molecules by the electron beam is indeed high when compared with the activation energies required in normal chemical reactions (a few kilocalories) and might be expected to result in random bond cleavages. However, despite the high energy, there is a specific chemistry behind the fragmentation behavior of most organic compounds. It is the set of rules and reactions constituting this chemistry that enables us to use mass spectrometry as a tool for the elucidation of organic molecular structures. One can consider the fragments formed in the mass spectrometer in much the same way that chemical degradation products are used for structural analyses; in this case, however, the degradation is carried out under high-vacuum conditions using energetic electrons as the reagent. This section provides an introduction to the chemistry of EI mass spectrometry and is followed by discussions of the techniques for using mass spectrometry in various types of problems.

Nearly every mass spectral reaction can be placed into one of two major categories

1. Simple cleavage
2. Multicenter fragmentations

SIMPLE CLEAVAGE

Simple cleavage is the fragmentation of a single covalent bond to give an ion and a neutral particle.

Equation 2.18 shows six possible ionic products that can be formed by simple cleavage of bonds in the molecular ion of a tetra-atomic molecule.

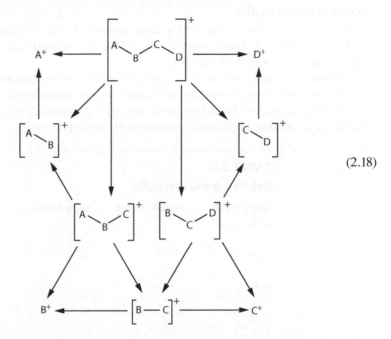

(2.18)

These primary products may also fragment by simple cleavage to give a "mass spectrum" now containing 10 ions, including the molecular ion.

Note also that some of the ions may be formed by more than one pathway. This is a prime complicating factor in studying mass spectral fragmentations but can be recognized by a number of methods that will soon become evident. Fortunately, although every bond is susceptible to fragmentation, the probability for cleavage is greater for some than for others. Those cleavages of high probability result in the more intense peaks in a spectrum, and are therefore important in the analysis of mass spectral data for structural information.

Simple cleavages of a molecular ion could lead to a rather complex mass spectrum (10 ions for a tetra-atomic molecule; 212 ions for an arrangement of 20 atoms). In reality, certain bonds are more likely to break than others, a phenomenon that leads to the formation of relatively few abundant ions.

Three related factors determine which bonds in a molecular ion will break and which ions will be formed in predominance.

1. Relative bond strengths
2. Stability of the ions formed
3. Neutral fragment stability

Ion stability is the most important of these factors, although there is a complex inter-action between all three forces that makes it difficult to predict the one that will determine the fragmentation course for any given molecule; there may even be two or more parallel fragmentations, each directed by a different driving force. Although the three factors cannot truly be separated, some comments about each are useful.

Relative Bond Strengths

Relative bond strengths of some of the more common bonds in organic molecules are shown in Table 2.6. Bond strength ranges from the weakest in the upper left to strongest in the lower right of the table.

The strongest bonds are those of the triple and double variety, and we might expect them to withstand the high energies better than the corresponding single bonds. In fact, this is the case; there is seldom any evidence for the cleavage of a multiple bond when there are weaker single bonds present in the molecular ion.

TABLE 2.6
Relative Bond Strengths

Single Bonds	Double Bonds	Triple Bonds
N–N		
C–I		
N–O		
S–S		
C–S		
C–Si		
C–Br		
C–N		
C–Cl		
C–C		
S–H		
C–O		
O–Si		
N–H		
C–H		
O–H		
C–F		
	N=N	
	C=S[a]	
	N=O	
	C=C	
	C=N	
	C=O[b]	
		C≡C
		C≡N

[a] For CS_2.
[b] For ketones.

Among the weakest of bonds listed in this table are those between carbon and the heavier halogen atoms. This fact is experimentally borne out in the mass spectrometer by the observation that the carbon-halogen bond of alkyl bromides and iodides is easily broken, whereas that of alkyl chlorides is less susceptible to fragmentation. Figure 2.11 shows the mass spectra of three butyl halides.

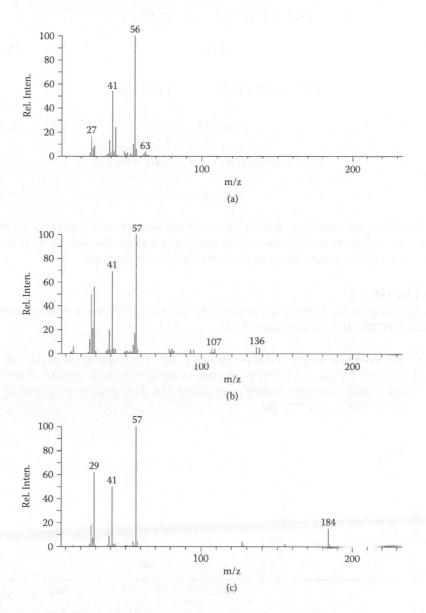

FIGURE 2.11   Mass spectra of (a) n-butyl chloride, (b) n-butyl bromide, and (c) n-butyl iodide.

The major fragmentation of 1-iodobutane and 1-bromobutane is loss of the halogen atom by simple cleavage to give the even-electron $C_4H_9^+$ ion of mass 57 (Equations 2.19 and 2.20). Cleavage of the carbon-chlorine bond in 1-chlorobutane is more difficult, and other fragmentations result, most notably elimination of HCl (a multicenter fragmentation discussed further elsewhere) to give the $C_4H_8^+$ ion at m/z 56 (Equation 2.21).

$$CH_3CH_2CH_2CH_2I \quad \xrightarrow{e^-} \quad C_4H_9^+ + I^\circ$$

$$m/z\ 57 \tag{2.19}$$

$$CH_3CH_2CH_2CH_2Br \quad \xrightarrow{e^-} \quad C_4H_9^+ + Br^\circ$$

$$m/z\ 57 \tag{2.20}$$

$$CH_3CH_2CH_2CH_2Cl \quad \xrightarrow{e^-} \quad C_4H_8^+ + HCl$$

$$m/z\ 56 \tag{2.21}$$

Such straightforward correlations between bond strengths and fragmentation patterns are seldom observed, but it will be useful to glance at this table as we discuss various types of fragmentations and various classes of compounds.

EXERCISE 2.4

Assign compositions to the major peaks in the spectrum of fluorochlorobromomethane, CHFClBr ($M = 146$) in Figure 2.12.

- The molecular weight of this compound is 146. The composition of this ion is $^{12}C^{19}F^{35}Cl^{79}Br$, or $CF^{35}Cl^{79}Br$. A peak of medium intensity appears at this mass, with a stronger isotope peak at m/z 148. This peak is a mixture of $CF^{37}Cl^{79}Br$ and $CF^{35}Cl^{81}Br$.

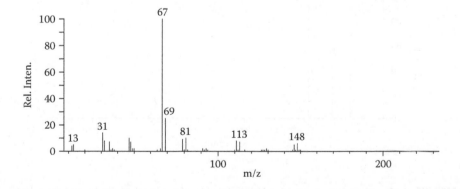

FIGURE 2.12   Mass spectrum of fluorochlorobromomethane.

- When the mass of a fluorine atom (19 Da) is subtracted from this, we obtain 127, and there is indeed a small m/z 127 peak ($C^{35}Cl^{79}Br$), with a stronger isotope peak at m/z 129 ($C^{37}Cl^{79}Br$ and $C^{35}Cl^{81}Br$).
- Doing the same for chlorine and bromine (35 and 79 Da, respectively), we arrive at masses that correspond to the peaks at m/z 111 ($CF^{79}Br$ and $CF^{81}Br$) and 67 ($CF^{35}Cl$ and $CF^{37}Cl$).
- Note the relative intensities of these three fragment peaks:
  - The bromine radical is lost most easily, giving the most intense peak in the spectrum (m/z 67).
  - Loss of the chlorine atom leads to a smaller peak (m/z 111).
  - Cleavage of the carbon-fluorine bond occurs only with reluctance (m/z 127).

Compare the isotope peak intensities with those in Table 2.4.

Stability of Ions

Discussing the stability of ions formed in the mass spectrometer can best be done by referring to the large body of knowledge concerning the solution reactions of organic compounds. Remembering that we are concerned mainly with positively charged ions in the mass spectrometer, and those factors that promote carbonium ion reactions will be of prime importance. This is clearly demonstrated in Figures 2.13a and

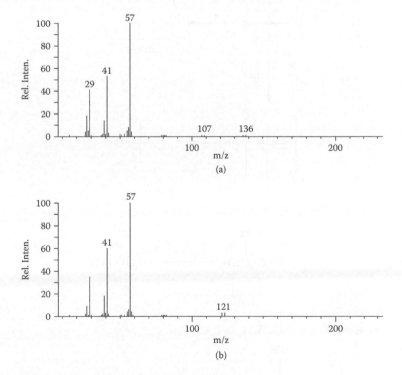

FIGURE 2.13   Mass spectra of (a) 2-bromobutane and (b) *tert*-butyl bromide.

2.13b, the mass spectra of the two isomers of 1-bromobutane (Figure 2.11b). From the relative peak intensities (compare the M−Br peak intensities with the intensities of the molecular ions), we see that cleavage of the carbon-bromine bond in *tert*-butyl bromide occurs most readily to give the stable *tert*-butyl cation—no molecular ion is observed at all in this spectrum. The corresponding cleavages in 2-bromobutane and 1-bromobutane give progressively weaker peaks, and the elimination of HBr to give the even-electron $C_4H_8^+$ ion (*m/z* 56) becomes a significant competing fragmentation pathway.

It should be noted that the high-energy M−Br ions formed in the mass spectrometer no doubt rearrange to some common ion before leaving the source. Nonetheless, the activation energy for the fragmentation, which determines the rate of the reaction and thus the intensities of the peaks, is dependent on the structure of the initially formed ion, no matter how fleeting its existence.

It is significant that this series of compounds behaves in an analogous manner under solvolytic conditions. That is, *tert*-butyl bromide reacts rapidly under S_N1 conditions first to form the tri-methyl carbonium ion, then the product *tert*-butanol (Equation 2.22), while 1-bromobutane reacts slowly under these conditions, and the chief product is not the alcohol but a mixture of butenes (Equation 2.24). 2-Bromobutane is intermediate in its reactivity (Equation 2.23).

(2.22)

(2.23)

$$CH_3CH_2CH_2CH_2\text{-Br} \xrightarrow{\ e^-\ } C_4H_9^+ \text{ and } C_4H_8^+$$

$$m/z\ 57 \qquad m/z\ 56$$

$$\downarrow H_3O^+$$

(2.24)

Butenes

To carry this analogy a little further, allylic and benzylic halides fragment faster than the saturated counterparts. The carbonium ions formed in both types of reaction are stabilized by resonance with neighboring π-electron systems (Equation 2.25). Thus, the carbon-halogen bonds in the unsaturated compounds are more susceptible to both solvolytic and EI-induced cleavage.

$$X^- \text{ (in solvolysis)}$$
$$+$$
$$X^{\bullet} \text{(in m.s.)}$$

(2.25)

The correlations presented here cannot of course be considered quantitative since the reaction processes are quite different. Reactions in solution depend on such factors as solvation and equilibria, whereas mass spectral reactions are strictly unimolecular decompositions of activated molecular ions. However, as a qualitative approach, these comparisons can be useful in the evaluation of spectra of related compounds. Differences in the intensities of various fragment peaks have been used to distinguish between numerous sets of isomeric compounds.

Nonbonding electrons of heteroatoms also add notable stability to a carbonium ion.

The bromination of alkenes, which proceeds through the carbonium ion a (the bromonium ion), is representative of this type of stabilization in chemical reactions.

$$R\text{—}CH=CH\text{—}R + Br_2 \longrightarrow R\text{—}\overset{\displaystyle CH\text{—}CH}{\underset{\displaystyle \underset{a}{Br}}{\diagdown\overset{+}{}\diagup}}\text{—}R + Br$$

(2.26)

$$\downarrow$$

$$R\text{—}\underset{\underset{Br}{|}}{CH}\text{—}\underset{\underset{Br}{|}}{CH}\text{—}R$$

In mass spectrometry the cyclic ion b has been proposed to account for its abnormally high abundance with respect to other ions in the homologous series. While the $C_4H_8X^+$ (where X is chlorine or bromine) ion is often a very intense peak in

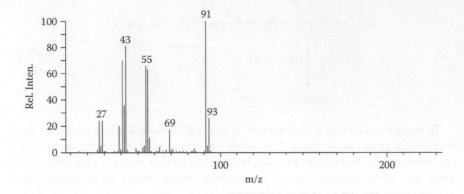

FIGURE 2.14 Mass spectrum of 1-chlorohexane.

the spectra of *n*-alkyl halides, the $C_3H_6X^+$ and $C_5H_{10}X^+$ ions are negligible (as in Figure 2.14, where *m/z* 91 corresponds to C_4H_8Cl, or $M-C_2H_5$).

$$\text{(cyclic structure labeled } X^+ \text{, } b)$$

More typically, the nonbonding electrons of heteroatoms stabilize a charge on the carbon bearing the heteroatom itself.

Termed *alpha* (α) *cleavage*, the fragmentation leading to this type of ion is exceedingly common in all classes of organic compounds. The fragmentations of alcohols, ethers, amines, and sulfides are all highly directed by this type of positive charge stabilization (Equations 2.27 to 2.32).

$$R\!-\!CH_2\ \overset{+X}{\cdots}\ CH_2\ \longrightarrow\ H_2C\overset{X^+}{\diagup}\!CH_2 + R^\bullet \tag{2.27}$$

$$X = Br,\ m/z\ 135$$
$$X = Cl,\ m/z\ 91$$

$$R\!-\!CH\!-\!R'\ \longrightarrow\ R\!-\!CH \quad \text{and/or} \quad CH\!-\!R' \tag{2.28}$$
$$\underset{\overset{+}{\bullet}OH}{}\qquad\qquad \underset{\overset{+}{}OH}{} \qquad\qquad \underset{\overset{+}{}OH}{}$$

$$R\!-\!CH_2\overset{+}{O}\!-\!CH_2R'\ \longrightarrow\ R\!-\!CH_2\overset{+}{O}\!=\!CH_2 \quad \text{and/or} \quad H_2C\!=\!\overset{+}{O}CH_2R' \tag{2.29}$$

$$R\!-\!\underset{\underset{\displaystyle\cdot SH}{\overset{+}{|}}}{CH}\!-\!R' \longrightarrow R\!-\!\underset{\overset{+}{\|}SH}{CH} \quad\text{and/or}\quad HC\!-\!R' \atop \overset{+}{\|}SH \tag{2.30}$$

$$R\!-\!\underset{\underset{\displaystyle\cdot NH_2}{\overset{+}{|}}}{CH}\!-\!R' \longrightarrow R\!-\!\underset{^+NH_2}{CH} \quad\text{and/or}\quad HC\!-\!R' \atop ^+NH_2 \tag{2.31}$$

$$R\!-\!\underset{\underset{\displaystyle\cdot NHR''}{\overset{+}{|}}}{CH}\!-\!R' \longrightarrow R\!-\!\underset{^+NHR''}{CH} \quad\text{and/or}\quad HC\!-\!R' \atop ^+NHR'' \tag{2.32}$$

Likewise, unsaturated compounds such as ketones and aldehydes, esters, and amides fragment in the same manner. The spectrum of methyl ethyl ketone (2-butanone) in Figure 2.15 is representative of α-cleavage. The two most intense peaks in the spectrum, at m/z 43 and 57, are formed by cleavage on either side of the carbonyl group (Equation 2.33).

$$\text{CH}_3\!-\!\overset{\overset{\displaystyle O}{\|}}{C}\!-\!\text{C}_2\text{H}_5 \tag{2.33}$$

$$\text{CH}_3\!-\!\overset{+}{C}\!\equiv\!O + \overset{\cdot}{\text{C}_2\text{H}_5} \qquad \text{C}_2\text{H}_5\!-\!\overset{+}{C}\!\equiv\!O + \overset{\cdot}{\text{CH}_3}$$
$$\text{m/z } 43 \qquad\qquad\qquad \text{m/z } 57$$

$$\text{CH}_3\!-\!\text{COCl} + \text{AlCl}_3 \longrightarrow \text{CH}_3\!-\!\text{CO}^+ \cdots \text{AlCl}_4^- \tag{2.34}$$

The ions described are not so familiar in normal chemical reactions, but the potential acylium ion formed from acetyl chloride and aluminum chloride (Equation 2.34) might be considered analogous to the mass spectral fragment ions.

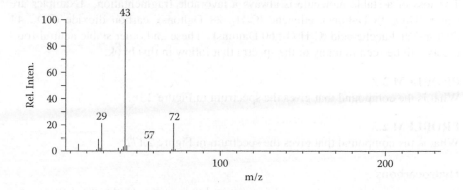

FIGURE 2.15 Mass spectrum of 2-butanone.

Further, the stabilities of acylium (and other α-cleavage products) ions are not surprising since they are isoelectronic with the corresponding stable cyano molecules R—CN.

$$CH_3 — C \equiv O^+ \qquad CH_3 — C \equiv N:$$

$$C_2H_5 — C \equiv O^+ \qquad C_2H_5 — C \equiv N:$$

Another example is mesitoic acid, which is esterified through the mesitoyl ion when treated with concentrated sulfuric acid in alcohol (Equation 2.35).

(2.35)

Neutral Fragment Stability

The loss of a stable molecule is always a favorable fragmentation. Examples are water (H_2O, 18 Daltons), ethylene (C_2H_4, 28 Daltons), carbon dioxide (CO_2, 44 Daltons) and acetic acid ($C_2H_4O_2$, 60 Daltons). These and other stable neutral fragments will be seen in many of the spectra that follow in this book.

PROBLEM 2.2

What is the compound that gives the spectrum in Figure 2.16?

PROBLEM 2.3

What is the compound that gives the spectrum in Figure 2.17?

Hydrocarbons

Remembering that secondary and tertiary carbonium ions (and radicals) are more stable than primary, observe the mass spectrum of 2,6-dimethyloctane in Figure 2.18.

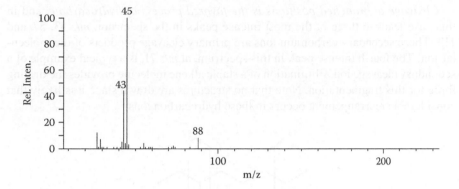

FIGURE 2.16 Problem 2.2.

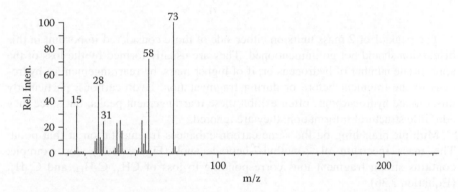

FIGURE 2.17 Problem 2.3.

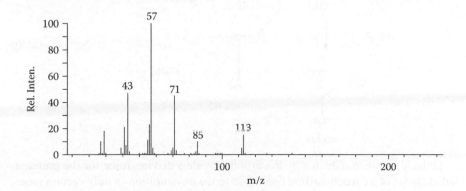

FIGURE 2.18 Mass spectrum of 2,6-dimethyloctane.

Cleavage at branched positions is the favored process of hydrocarbons and in this case leads to three of the most intense peaks in the spectrum, *m/z* 43, 57, and 113. These secondary carbonium ions are primary cleavage products of the molecular ion. The fourth intense peak in this spectrum, at *m/z* 71, is a typical example of a secondary cleavage ion. Elimination of a stable alkene molecule provides the driving force for this fragmentation. Note that no structures are drawn since it is likely that considerable rearrangement occurs in these hydrocarbon ions.

The peaks 1 or 2 mass units on either side of those considered important in this discussion should not go unmentioned. They are usually formed by the loss of the appropriate number of hydrogens or, if of higher mass, by rearrangement of hydrogens to the fragment before or during fragmentation. Hydrocarbons, particularly unsaturated hydrocarbons, often exhibit these rearrangement peaks, but since they add little structural information, they are ignored.

Multiple branching on the same carbon enhances fragmentation at that point. The mass spectrum of 7-methyl-7-heptadecanol (Figure 2.19), for example, contains strong fragment ions corresponding to loss of CH_3, C_6H_{13}, and $C_{10}H_{21}$ (Equation 2.36).

(2.36)

Isolated double bonds in a hydrocarbon provide a driving force for the preferential cleavage of a carbon-carbon bond *beta* to the unsaturation in only certain cases. Normally, rapid rearrangement of hydrogens within the molecular ions of olefins occurs faster than allylic cleavage; thus, there is actually a mixture of molecular

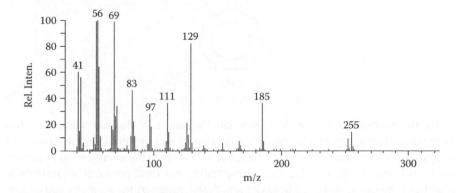

FIGURE 2.19 Mass spectrum of 7-methyl-7-heptadecanol.

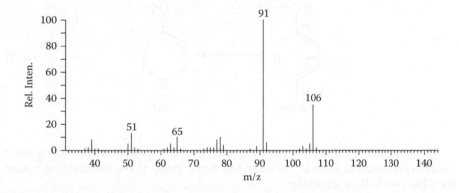

FIGURE 2.20 Mass spectrum of ethylbenzene.

ions formed immediately after ionization. Most notably, however, aromatic hydrocarbons containing alkyl substituents have a strong tendency to break a benzylic bond. The mass spectrum of ethylbenzene in Figure 2.20 has a very intense $M-CH_3$ ion at m/z 91. An intense peak at this mass (or at a mass corresponding to higher homologs of $C_7H_7^+$: 105, 119, 133, ...) is often indicative of a substituted benzene ring (Equation 2.37).

$$CH_2 \overset{\nearrow}{\frown} CH_3 \longrightarrow C_7H_7^+ + CH_3^\bullet$$
$$m/z\ 91 \tag{2.37}$$

Although in most cases the structures of ions are only speculative, and drawn merely to explain significant fragmentation reactions, the unusual intensity of the $C_7H_7^+$ ion in many spectra has aroused the interest of several workers.

$$C_7H_7^+$$

By the appropriate use of deuterium labeling and low electron voltages, it has been shown that a ring enlargement rapidly takes place to form the highly stabilized tropylium ion; this ion is aromatic in character. Whether ring enlargement is universal is still open for debate, for the corresponding ion from para-methoxyethylbenzene has been shown to be the benzylic ion. Stabilization by the methoxy substituent apparently prevents ring enlargement in this case (Equation 2.38).

(2.38)

The remainder of this book presents both types of ions without distinction since either representation leads to the same functional group in the molecular structure for which we will be searching.

Since nonaromatic compounds do not usually exhibit strong peaks corresponding to cleavage of an allylic bond, the locations of double bonds in alkenes is difficult. However, in specific cases such as myrcene, which has a doubly allylic bond, cleavage may be highly favored. An intense $C_5H_9^+$ ion *(m/z* 69) is formed from myrcene (Equation 2.39), but it is not formed from the isomeric compound allo-ocimene (Equation 2.40).

$$C_5H_9^+ + C_5H_7^{\cdot}$$
$$m/z\,69$$

Myrcene

(2.39)

$$C_5H_9^+$$

Allo-ocimene

(2.40)

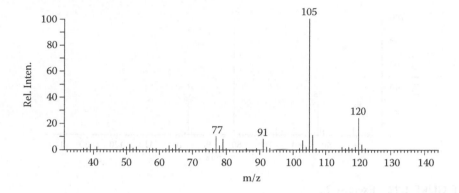

FIGURE 2.21 Exercise 2.5.

EXERCISE 2.5

What are the compounds that give the spectra in Figures 2.21 and 2.22?

- The highest mass ion in Figure 2.21 does not fail any of the tests for a molecular ion, so we can assume that the molecular weight is 120.
- The only fragment ion of any significance is at m/z 105, or M–CH₃, and the spectrum is very similar in appearance to that in Figure 2.20.
- An upward shift in mass of 14 units leads to the structure methylethylbenzene.
- Mass spectrometry in this case is useless in specifying the relative positions of the two alkyl functions on the ring (*ortho*, *meta*, or *para*). Isopropylbenzene gives a similar spectrum (see Equation 2.41).

$$\longrightarrow \quad C_8H_9^+ + CH_3^{\bullet} \qquad (2.41)$$
$$m/z\ 105$$

- The spectrum in Figure 2.22 is similar in appearance to the two spectra preceding and could therefore be a substituted aromatic compound.
- The loss of a methyl radical is characteristic of an ethyl side chain (as previously discussed), which leaves only 67 mass units for the remainder of the molecule.
- Since nitrogen is excluded on the basis of an even molecular weight, the only possible composition that adds up to 67 Da is C_4H_3O, which corresponds to the furan nucleus.
- The compound is then ethylfuran, but without reference spectra, the position of the alkyl function again cannot be stated.

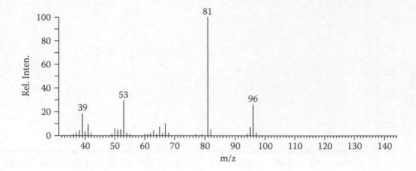

FIGURE 2.22 Exercise 2.5.

- By analogy with the formation of the tropylium ion, ring expansion may occur in the $M–CH_3$ ion to form an analog of benzene (Equation 2.42). Sufficient studies have not been performed to firmly establish such a rearrangement.

$$\text{(2.42)}$$

m/z 81

General Compounds

Simple cleavage becomes an even more important process in molecules containing heteroatoms such as nitrogen and oxygen.

Table 2.7 illustrates some of the more common ions formed by simple cleavage α to the heteroatoms in alcohols, ketones, esters, amines, and sulfur-containing compounds. The significance of these ions was evident when we examined the spectrum of 2-butanone (Figure 2.15). The structure was completely described by the two most intense fragment peaks in the spectrum.

Alcohols behave in an analogous manner, as shown by the mass spectrum of 2-butanol in Figure 2.23. The three possible α-cleavage products compose a large portion of the total ion current in the spectrum (Equation 2.43). Only a small molecular ion is observed.

$$\text{(2.43)}$$

TABLE 2.7
Ions Formed by Simple Cleavage

Ion	Nominal Mass	Origin
H–C≡O⁺	29	Aldehydes
H₂C=NH₂⁺	30	Amines
H₂C=OH⁺	31	Alcohols
CH₃C≡O⁺	43	Ketones
H₂C=SH⁺	47	Sulfides
H₂C=Cl⁺	49	Chloro compounds
CH₃O–C≡O⁺	59	Methyl esters
CHCl=Cl⁺	83	Chloro compounds
CH₃OCO——CH ‖ ⁺NH₂	88	Amino acid esters
[phenyl]=C=O⁺	105	Benzoyl compounds

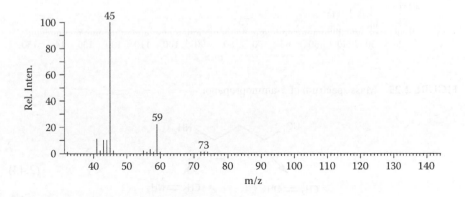

FIGURE 2.23   Mass spectrum of 2-butanol.

PROBLEM 2.4
What is the compound that gives the spectrum in Figure 2.24?

Likewise, nitrogen-containing compounds almost always give strong fragment peaks arising by α-cleavage. Nitrogen promotes fragmentation more than oxygen (Equation 2.44), as shown by the spectrum of 3-aminopropanol in Figure 2.25.

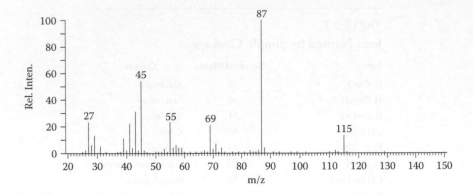

FIGURE 2.24   Problem 2.4.

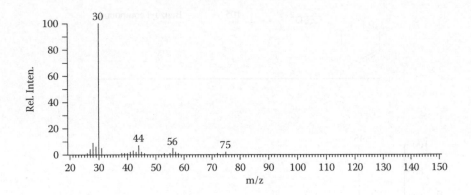

FIGURE 2.25   Mass spectrum of 3-aminopropanol.

$$HO \diagup\diagdown\diagup NH_2$$

$$CH_2 \!=\! \overset{+}{O}H \qquad\qquad CH_2 \!=\! \overset{+}{N}H_2$$ (2.44)

$$m/z\ 31 \qquad\qquad\qquad m/z\ 30$$

The nitrogen-containing ion at m/z 30 ($CH_2\!=\!NH_2^+$) is about 10 times more abundant than the corresponding oxygen-containing ion at m/z 31 ($CH_2\!=\!OH^+$).

As might be suspected from the simplicity of the examples presented to this point, simple cleavage does not go far in describing the mass spectra of organic compounds. When molecules become larger and more complex, the functional groups begin to interact so that more complicated fragmentation reactions occur. But, before examining the mass spectra of the more complex molecules, let us look at some of the common multicenter fragmentation reactions. It must once again be stressed that the ion structures and the mechanisms leading to them are only for instructional purposes and do not try to represent the actual structures and processes.

MULTICENTER FRAGMENTATIONS

Multicenter fragmentations encompass eliminations, retro Diels-Alder, rearrangement, and other reactions involving the cleavage of more than one bond and the concurrent formation of one or more new bonds.

In the linear tetra-atomic molecule of *ABCD*, a multicenter fragmentation might lead to yet another ion, with a new bond joining atoms *A* and *D* (Equation 2.45). Not recognizing the occurrence of such fragmentation reactions might lead to the false conclusion that *A* and *D* are joined in the original structure. Like simple cleavage reactions, however, these more complicated reactions can be anticipated and utilized in analyzing mass spectral data for structural evidence.

$$\underset{B-C}{\overset{A\quad D}{\bigvee}}H \xrightarrow{e^-} A-D^+ + B{=\!=}C \qquad (2.45)$$

Multicenter fragmentations encompass a wide variety of reactions. To study each of these in detail would require many pages in several chapters, with much of the material repetitious. Instead, it is useful when introducing mass spectral fragmentations first to discuss the general relationships of the more common reactions. The details of each can subsequently be analyzed in the course of specific interpretation problems.

The transfer of a hydrogen atom from one atom to another within an ion is one of the most common fragmentation processes occurring in the mass spectrometer. The elimination of water from alcohols (Figure 2.19) involves the transfer of a hydrogen atom from a carbon atom to an oxygen atom. In a similar manner, alkyl amines, esters, sulfides, and halides undergo elimination reactions. Other hydrogen migrations occur to give rearranged ions (olefinic hydrocarbon and aromatic system, which lead to the tropylium ion); still others proceed with a high degree of positional specificity to give ions characteristic of certain structural features.

Eliminations

The general elimination fragmentation can be depicted as the process that follows, where *X* is described in Table 2.8. The number of carbon atoms between the departing groups (hydrogen and the *X* function) may vary and has been shown to be specific in several instances. That is, a hydrogen atom is abstracted from a carbon that is a specific number of atoms removed from the *X* group (Equation 2.46).

$$\underset{X}{\overset{H}{\bigvee}}(CH_2)n \xrightarrow{e^-} \left[\triangleright (CH_2)n \right]^+ + HX \qquad (2.46)$$

TABLE 2.8

Neutral Molecules Eliminated in the Mass Spectrometer

Molecule	Mass
NH_3	17
H_2O	18
CH_3OH	32
H_2S	34
HCl	36
C_2H_5OH	46
CH_3COOH	60
HBr	80
HI	128

Alcohols

The elimination of water is the most thoroughly documented EI-induced reaction of this class of compounds and is recognized by a peak 18 mass units lower than the molecular ion. The attention accorded this reaction has revealed it to be significantly different from the corresponding thermal or acid-catalyzed reaction. Whereas the normal chemical elimination of water from acyclic alcohols involves adjacent (1,2)-functions, a six-membered transition state is preferred in the electron-bombardment-induced reaction. Thus, a hydrogen is abstracted from a carbon four atoms away from that bearing the hydroxyl group, making this a 1,4-elimination (Equation 2.47).

$$R \longrightarrow CH_2 - CH_2 - CH_2 - CH_2 - OH$$

(2.47)

Although the M–H_2O ion is represented here by a cyclobutane structure, this need not be its actual structure; no evidence is available to confirm such structures, but it is felt that new bond formation in the product ion lowers the energy requirements of the reaction. The fact that hydrogen is abstracted from the δ-position has been conclusively demonstrated by extensive deuterium-labeling studies. That is, replacement of the hydrogens on the δ-carbon by deuterium results in an almost-exclusive elimination of HDO, whereas deuteriums on the α-, β-, or γ-carbon atoms remain on the fragment ion with the loss of H_2O.

Through what is most certainly a different transition state, the elimination of water is sometimes accompanied by the expulsion of a molecule of ethylene. An M–46 peak thereby results. The transition state in Equation 2.48 may be postulated, although the deuterium-labeling studies to confirm the location from which the hydrogen comes have not been performed.

$$R-CH \underset{\overset{|}{H}}{\overset{CH_2 - CH_2}{\diagdown}} \begin{array}{c} \\ CH_2 \\ \overset{+}{O} \\ | \\ H \end{array} \longrightarrow R - \overset{+}{C}H_2 - \overset{\bullet}{C}H_2 + H_2O + C_2H_4 \qquad (2.48)$$

EXERCISE 2.6

Use the information just discussed to match the two spectra in Figures 2.26 and 2.27 with the following structures for 1-hexanol and 2-methyl-1-pentanol.

$$CH_3CH_2CH_2CH_2CH_2CH_2OH$$
1-Hexanol

$$\overset{\overset{\textstyle CH_3}{|}}{CH_3CH_2CH_2CHCH_2OH}$$
2-Methyl-1-pentanol

- Figure 2.26 exhibits a very weak, but significant, molecular ion peak at 102 Da.
- Fairly abundant ions appear at m/z 84 in both spectra corresponding to M–H_2O species.
- An M–31 ($CH_3O^{\bullet}$) ion at m/z 71 is in Figure 2.27, but this does not a priori distinguish between the two compounds.

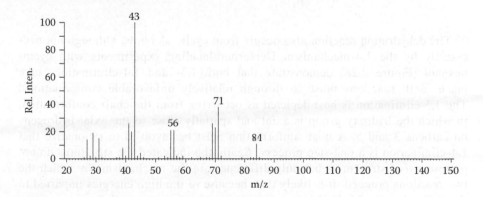

FIGURE 2.26 Exercise 2.6.

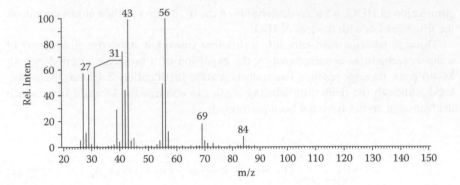

FIGURE 2.27   Exercise 2.6.

• The main difference between the two spectra is the intense peak at *m/z* 56
  in Figure 2.27, which is nearly absent from the other. This ion is M–46
  (M–H$_2$O–C$_2$H$_4$) and must therefore arise from 1-hexanol in the manner
  described (Equation 2.49), while 2-methyl-1-pentanol does not undergo this
  rearrangement fragmentation (Equation 2.50).

$$CH_3CH_2CHCH_2CH_2CH_2 \longrightarrow [CH_3CH_2CH \!=\!\!=\! CH_2]^+ \qquad (2.49)$$
$$m/z\ 56$$

$$CH_3CHCH_2CHCH_2 \;/\!/\!\!\longrightarrow\; [CH_3CH \!=\!\!=\! CH_2]^+ \qquad (2.50)$$
$$CH_3 \qquad\qquad m/z\ 42$$

The dehydration reaction also occurs from cyclic alcohols, although not necessarily by the 1,4-mechanism. Deuterium-labeling experiments with cyclohexanol (Figure 2.28) demonstrate that both 1,3- and 1,4-eliminations take
place. Both reactions must go through relatively unfavorable conformations.
The 1,3-elimination is best depicted as occurring from the chair conformation,
in which the hydroxy group is axial and spatially close to the axial hydrogens
on carbons 3 and 5. A boat conformation must be involved to rationalize that
1,4-elimination is a one-step process. Again, the indicated ion structures do not
represent true structures, but only differences in the mechanisms by which the
two reactions proceed. It is likely that, because of the high energies imparted to
the molecules, the M–H$_2$O ions eventually (perhaps even rapidly) rearrange to a
common ion (Equation 2.51).

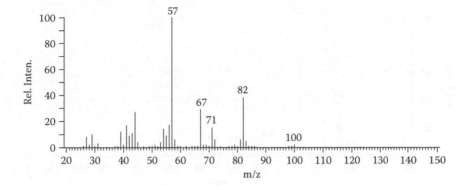

FIGURE 2.28   Mass spectrum of cyclohexanol.

$$\qquad\qquad\qquad\qquad\qquad\qquad\qquad\qquad (2.51)$$

The deuterium-labeling experiments indicate that 1,4-eliminations are favored, a fact that is supported by the spectra of the isomeric 4-*tert*-butylcyclohexanols. Contrary to the previously held belief that a more crowded (generally axial) hydroxy group will be eliminated faster, *cis*-4-*tert*-butylcyclohexanol eliminates water more slowly than the *trans* isomer (Equation 2.52). This would appear to indicate that 1,4-eliminations from a boat conformation are preferred, and that the elimination is indeed *cis*.

$$\qquad\qquad\qquad\qquad\qquad\qquad\qquad\qquad (2.52)$$

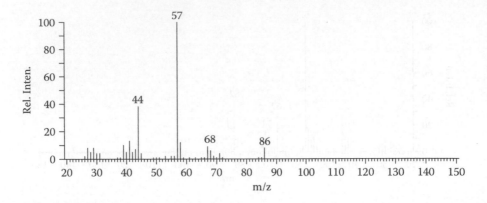

FIGURE 2.29   Mass spectrum of cyclopentanol.

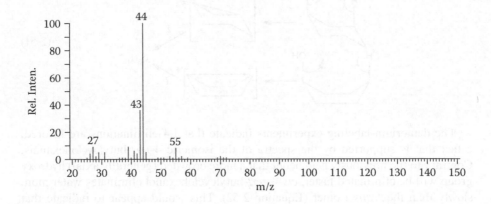

FIGURE 2.30   Mass spectrum of cyclobutanol.

Furthermore, the mass spectra of cyclopentanol (M = 86) and cyclobutanol (M = 72) (Figures 2.29 and 2.30, respectively) exhibit progressively weaker M–H_2O peaks. Both compounds must go through a more strained five-membered transition state to eliminate water.

Halides

Rather surprisingly, if the hydroxy group is replaced by a halogen atom, elimination of HX (X = Cl or Br) appears to prefer a five-membered transition state (Equation 2.53). Deuterium-labeling studies with n-butyl and n-pentyl chloride demonstrate that abstraction of the hydrogen from the γ-carbon is the favored process. There are also indications that cyclic chlorides and bromides eliminate HX by the 1,3-mechanism.

$$R\!-\!CD_2CH_2CH_2Cl^+ \longrightarrow R\!-\!\overset{\overset{\displaystyle D}{|}}{CD}\underset{CH_2}{\overset{\overset{\displaystyle {}^+Cl}{|}}{\diagup}\!\!\!\diagdown CH_2}$$

$$\longrightarrow \left[\, R\!-\!CD\overset{\displaystyle CH_2}{\underset{\displaystyle CH_2}{\diagdown\!\!|\!\!\diagup}} \,\right]^{+} + \;DCl \qquad (2.53)$$

Other Compounds

Not so surprisingly, the elimination of acetic acid from acetate esters occurs in both 1,2- and 1,4 manners; six-membered transition states may be proposed for each type of elimination, making both fragmentation mechanisms reasonable (Equation 2.54).

$$R\!-\!\underset{\underset{\displaystyle H}{|}}{CH}CH_2CH_2\underset{\underset{\displaystyle {}^+OCOCH_3}{|}}{CH_2} \longrightarrow \left[\begin{array}{c} H_2C\!-\!CH_2 \\ |\qquad | \\ R\!-\!CH\!-\!CH_2 \end{array}\right]^{+}$$

$$+\;HOAc$$

$$(2.54)$$

$$R\!-\!CH_2CH_2\overset{\overset{\displaystyle H}{|}}{CH}\!-\!CH_2O\!-\!\overset{\overset{\displaystyle {}^+O}{\parallel}}{C}CH_3$$

$$\downarrow$$

$$\left[\,R\!-\!CH_2CH_2CH=\!CH_2\,\right]^{+} + HOAc$$

Little is known about the mechanisms of elimination from other classes of compounds. Deuterium labeling is probably the best technique for studying these reactions, but this involves considerable synthetic work, and such has not yet been reported.

Caution must be exercised when interpreting mass spectral eliminations in the manner discussed for one of the most common "artifacts" in mass spectrometry is sample decomposition, particularly by elimination of neutral molecules of those substances listed in Table 2.5 prior to volatilization and ionization. The high temperatures required to vaporize some materials are great enough to promote thermal elimination, which leads in most cases to mixtures of olefins. The occurrence of thermal elimination would certainly produce ambiguous results but is detectable by a number of methods. Low electron energies (~10 to 15 eV) usually produce less fragmentation, leaving the molecular ions as the most abundant species; a thermal elimination would therefore result in a "molecular ion" at the M–HX mass position. In addition, thermal dehydrations usually cause the spectrum to change with time as the analysis proceeds; the M–HX ion increases with respect to the true molecular ion. The problem of thermal decomposition is particularly critical when high-temperature reservoir inlets are used. The two mass spectra of cholesterol in Figures 2.31a and 2.31b demonstrate the differences that can arise under differing inlet conditions. The spectrum in Figure 2.31a was determined using a heated reservoir inlet at 250°C and exhibits a very intense M–H_2O peak at m/z 368. This elimination is mainly thermally induced since the other spectrum, obtained by introducing the sample on a direct

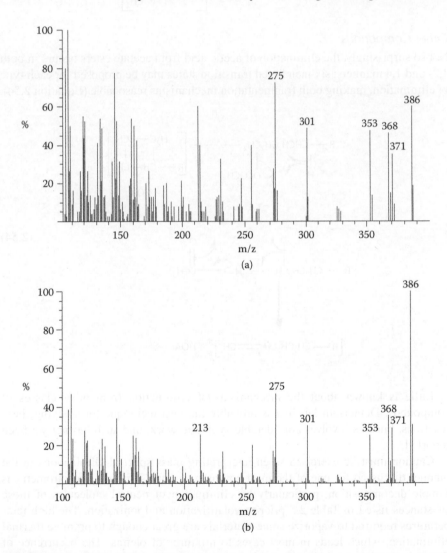

FIGURE 2.31 Mass spectra of cholesterol (a) determined through a heated reservoir inlet and (b) determined by direct insertion.

insertion probe at about 80°C, exhibits only a small M–H$_2$O peak. In addition, the latter spectrum exhibits considerably weaker fragment peaks in the low-mass region of the spectrum.

McLafferty Rearrangements

Perhaps the most thoroughly documented EI-induced hydrogen migration reaction is the McLafferty rearrangement. In the mass spectrum of 2-pentanone (Figure 2.32), a significant peak appears at m/z 58 corresponding to the loss of 28 mass units from the molecular ion. A peak at the same mass in the spectrum of 2-hexanone (Figure 2.33) corresponds to the odd-electron fragment formed by the loss of 42 mass units. The greater intensity of this peak in the latter spectrum indicates that secondary hydrogen is abstracted more readily than primary hydrogen. A six-membered transition state involving the transfer of a hydrogen from the position gamma (γ) to the carbonyl group leads to this ion in both spectra (Equation 2.55).

$$H_3C-\overset{\overset{\displaystyle O^+}{\|}}{C}-CH_2-CH_2-CH-R$$

(2.55)

$$\left[H_3C-\overset{\overset{\displaystyle OH}{|}}{C}=CH_2\right]^+ \; + C_2H_4 \quad \text{or} \quad C_3H_6$$

m/z 58

The other intense peaks in these spectra can be attributed to simple cleavages α to the carbonyl functions (Equations 2.56 and 2.57).

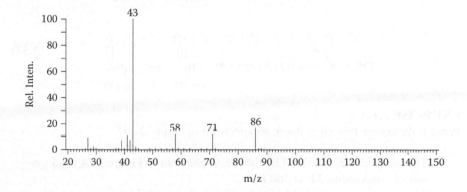

FIGURE 2.32   Mass spectrum of 2-pentanone.

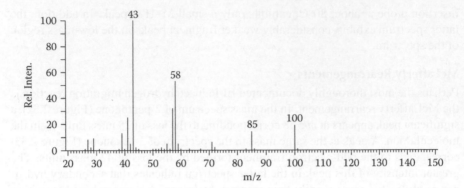

FIGURE 2.33　Mass spectrum of 2-hexanone.

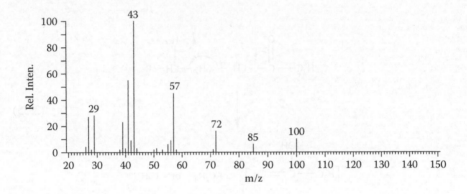

FIGURE 2.34　Exercise 2.7.

$$CH_3-\overset{\overset{\displaystyle O^+}{\|}}{C}-(CH_2)_2CH_3 \longrightarrow CH_3\overset{\overset{\displaystyle O^+}{\||}}{C} \text{ and } C_3H_7\overset{\overset{\displaystyle O^+}{\||}}{C} \qquad (2.56)$$
$$m/z\ 43 \qquad m/z\ 71$$

$$CH_3-\overset{\overset{\displaystyle O^+}{\|}}{C}-(CH_2)_3CH_3 \longrightarrow CH_3\overset{\overset{\displaystyle O^+}{\||}}{C} \text{ and } C_4H_9\overset{\overset{\displaystyle O^+}{\||}}{C} \qquad (2.57)$$
$$m/z\ 43 \qquad m/z\ 85$$

EXERCISE 2.7

What is the ketone that gives the mass spectrum in Figure 2.34?

- The compound represented by the mass spectrum in Figure 2.34 is an isomer of 2-hexanone, M$^+$ at 100 Da.
- The fragment at m/z 85 (M-15) identifies it as a methyl ketone.
- The other alkyl group must be a C_4H_9 unit.

- The peak at *m/z* 58 is no longer present but is replaced by another odd-electron (even-mass) ion at *m/z* 72 corresponding to M-28.
- The only solution to this problem, then, is 3-methyl-2-pentanone, which retains the methyl group when undergoing the McLafferty rearrangement (Equation 2.58).

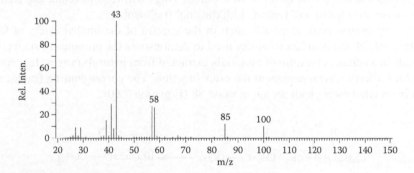

(2.58)

PROBLEMS 2.5 AND 2.6

The mass spectra in Figures 2.35 and 2.36 (Problems 2.5 and 2.6, respectively) are also of hexanone isomers. Solve.

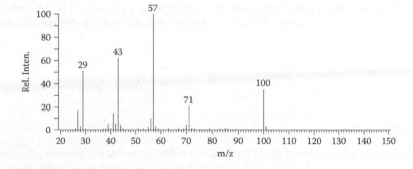

FIGURE 2.35 Problem 2.5.

FIGURE 2.36 Problem 2.6.

As demonstrated in the preceding examples, the McLafferty rearrangement may be recognized by the occurrence of odd-electron ions in the spectrum. A neutral olefin molecule is generally expelled, but in some cases, the charge remains with the olefinic fragment to expel a neutral enol molecule. Such is the situation with 1-phenyl-4-pentanone (Equation 2.59).

$$
\begin{array}{c}
\text{[structure: phenyl ring with } \overset{H}{\underset{|}{C}}H-CH_2-CH_2-\overset{\overset{O^+}{\|}}{C}-CH_3]
\longrightarrow
\left[\text{phenyl ring with } CH=CH_2\right]^+ + C_3H_6O \quad (2.59)
\end{array}
$$

m/z 104

Other rearrangement fragmentations that lead to odd-electron fragment ions are discussed further in the chapter.

Under normal circumstances, the charge will remain on the species that has the lower ionization potential. That is, the fragmentation will lead to the two fragments, charged and neutral, that have the lowest energy. Thus, highly substituted olefinic fragments and those conjugated with aromatic rings will tend to retain the charge. Otherwise, the charge will remain with the enol fragment.

A very intense peak is usually seen in the spectra of the methyl esters of fatty acids at *m/z* 74 and is in fact routinely used to demonstrate the presence of such compounds in a complex mixture of materials extracted from natural sources. Its genesis is a McLafferty rearrangement at the ester function. The corresponding fragmentation in an ethyl ester yields an ion of mass 88 (Equation 2.60).

$$
\underset{H}{\overset{\frown}{C_{14}H_{29}CH}}-CH_2\overset{\cdot}{-}CH_2-\overset{\overset{O^+}{\|}}{C}-OCH_3 \longrightarrow \left[H_2C=\overset{\overset{OH}{|}}{C}-OCH_3\right]^+ \quad (2.60)
$$

m/z 74

Deuterium-labeling studies have shown in this case, as with the ketones, that one hydrogen atom is specifically transferred from the γ-position. The γ,γ-dimethyl derivative of methyl nonanoate does not give the odd-electron ion (Equation 2.61).

$$
C_5H_{11}-\overset{\overset{CH_3}{|}}{\underset{\underset{CH_3}{|}}{C}}-CH_2-CH_2-\overset{\overset{O^+}{\|}}{C}-OCH_3 \;\; \cancel{\longrightarrow} \;\; m/z\ 74 \quad (2.61)
$$

Hydrogen migration may also take place to a doubly bonded atom other than oxygen. The aromatic π-electron systems of *n*-butylbenzene and phenetole (Figure 2.37) readily promote the hydrogen migration (Equations 2.62 and 2.63).

$$(2.62)$$

m/z 92

$$(2.63)$$

m/z 94

The steric limitations for this rearrangement are illustrated by the spectra of the two highly substituted aromatic compounds that follow. With one *ortho* methyl group, the rearrangement proceeds to give a peak at *m/z* 120, but if both *ortho* positions are substituted, the reaction is blocked (Equation 2.64).

$$(2.64)$$

m/z 134   m/z 120

Many other examples of McLafferty-like rearrangements will be encountered with various classes of compounds. In general, this rearrangement requires only a double bond and a hydrogen atom in the proper relative positions, as shown by Equation 2.65, where the letters represent various types of atoms. The hydrogen migration may occur before cleavage of the C–D bond, but it cannot, of course, occur afterward.

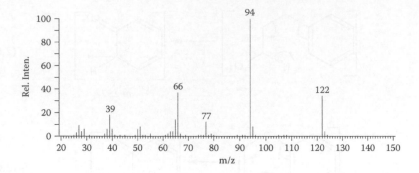

FIGURE 2.37 Mass spectrum of phenetole.

$$\text{(2.65)}$$

Some of the commonly observed fragments of McLafferty rearrangements of ketones and aldehydes, carboxylic acids, esters and amides, aromatic systems, and others are presented in Table 2.9. In all cases, the particular fragment mass must be considered along with homologs and related structures. For example, an ethyl ester gives the ion corresponding to $M{-}C_2H_4$ (Equation 2.66), and a methoxy-substituted aromatic ring would be expected to yield an ion of mass 122 (Equation 2.67). Note also that an ethyl ester can (and often does) undergo a McLafferty rearrangement in another direction.

$$R\!-\!\overset{\overset{O^{+}}{\|}}{C}\!-\!O\!-\!CH_2\!-\!CH_2 \longrightarrow [\,R\!-\!COOH\,]^{+} + C_2H_4 \qquad (2.66)$$

$$\qquad (2.67)$$

m/z 122

PROBLEM 2.7
Identify the compound that gives the mass spectrum in Figure 2.38.

Although the McLafferty rearrangement occurs in a wide variety of compounds, its full scope is not yet understood. Some seemingly minor changes in structure sometimes result in other fragmentations that are energetically more favorable.

TABLE 2.9
Fragments Formed by McLafferty Rearrangements

Fragment Ion

Nominal Mass

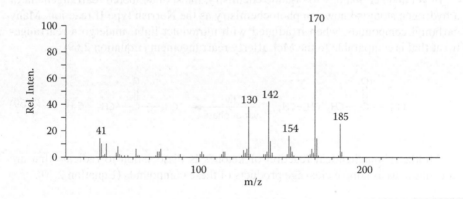

44

58

59

60

74

92

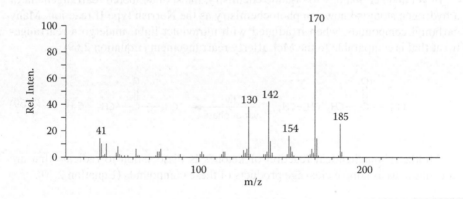

FIGURE 2.38 Problem 2.7.

Whereas *n*-butylbenzene and isobutylquinoline expel propylene, the loss of this even-electron fragment from isobutylnaphthalene is only a minor decomposition pathway (Figure 2.39). Only a small peak is observed at *m/z* 142 (M−42), while by far the most significant fragmentation is simple cleavage of the benzylic bond to lead to the base peak at *m/z* 141 (Equation 2.68).

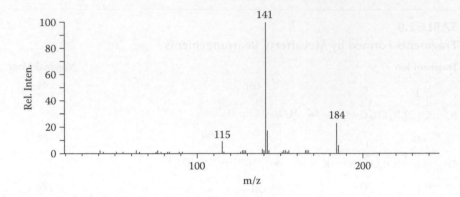

FIGURE 2.39   Mass spectrum of 1-isobutylnaphthalene.

$$\tag{2.68}$$

In yet another analogy to organic chemistry, the six-membered rearrangement of a hydrogen atom is known in photochemistry as the Norrish type II reaction. Many carbonyl compounds, when irradiated with ultraviolet light, undergo a rearrangement that is comparable to the McLafferty rearrangement (Equation 2.69).

$$CH_3 - \overset{\overset{O}{\|}}{C} - CH_2\text{-}CH_2\text{-}CH_3 \quad \xrightarrow[\text{vapor phase}]{hv} \quad CH_3 - \overset{\overset{O}{\|}}{C} - CH_3 + C_2H_4 \tag{2.69}$$

Another photochemical reaction of ketones is dissociation to radicals that are familiar to us as simple cleavage products of these compounds (Equation 2.70).

$$H_3C - \overset{\overset{O}{\|}}{C} - CH_3 \quad \xrightarrow[\text{vapor phase}]{hv} \quad H_3C - \overset{\overset{O}{\|}}{C}{}^\bullet + CH_3^\bullet \tag{2.70}$$

$$H_3C - \overset{\overset{O}{\|}}{C} - \overset{\overset{O}{\|}}{C} - CH_3$$

Retro Diels-Alder Fragmentations

The reverse Diels-Alder reaction is another multicenter fragmentation often observed in mass spectrometry. Cyclic olefins in particular undergo this reaction to expel a neutral olefinic fragment (Equations 2.71 and 2.72).

$$C_4H_6^+ + C_2H_4 \qquad (2.71)$$
$$m/z\ 54$$

$$+ \ C_2H_4 \qquad (2.72)$$
$$m/z\ 66$$

The spectrum of norbornene (Figure 2.40), for example, contrasts sharply with that of the saturated analog norbornane (Figure 2.41). The retro Diels-Alder fragmentation leads to the most intense peak in the norbornene spectrum at m/z 66. The spectrum of norbornane, on the other hand, is much more like that of a typical hydrocarbon, with even-electron (odd-mass) ions dominant.

Like the McLafferty rearrangements, retro Diels-Alder reactions occur in many types of compounds with various double-bond systems. The aromatic ring of tetralin, for example, promotes the expulsion of ethylene from its molecular ion (Equation 2.73).

$$+ \ C_2H_4 \qquad (2.73)$$
$$m/z\ 104$$

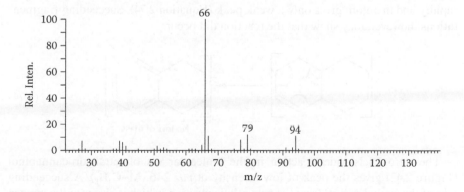

FIGURE 2.40   Mass spectrum of norbornene.

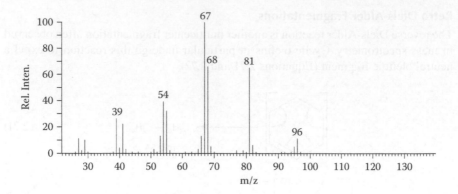

FIGURE 2.41   Mass spectrum of norbornane.

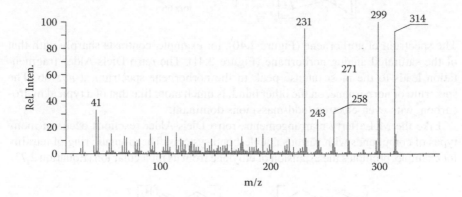

FIGURE 2.42   Mass spectrum of tetrahydrocannabinol.

In complex molecules containing several rings, this reaction may not be initially apparent due either to no loss of mass or to an unstable product that decomposes rapidly and therefore gives only a weak peak (Equation 2.74). Succeeding fragmentations, however, may show that the reaction did occur.

$$\left[\begin{array}{c}\end{array}\right]^{+} \longrightarrow \left[\begin{array}{c}\end{array}\right]^{+} \qquad (2.74)$$

No loss of mass

The retro Diels-Aider reaction in the molecular ion of tetrahydrocannabinol (Figure 2.42) gives the peak of low intensity at m/z 246 (M$-$C$_4$H$_8$). A succeeding loss of the methyl radical from the new allylic position (which is also α to the oxygen atom) results in a strong peak at m/z 231 (Equation 2.75).

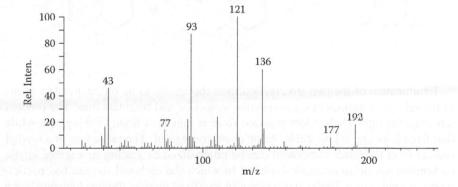

(2.75)

This example points out that, as with the McLafferty rearrangement, either fragment may retain the charge; the more highly substituted or conjugated fragment generally remains charged. In simple systems, this usually results in a charged diene fragment, but (as with tetrahydrocannabinol) the monoene may be the neutral species.

One additional example, the α- and β-ionones (Figures 2.43 and 2.44), points out how minor structural differences may affect the fragmentation course of molecules. The two compounds differ only in the position of the endocyclic double bond, but the spectra are quite different. The spectrum of α-ionone exhibits an intense peak at m/z 136, corresponding to the elimination of butylenes from the molecular ion. The β-ionone, on the other hand, has no peak at the mass expected for the elimination of ethylene; the only major fragmentation of the molecular ion is loss of a methyl radical (note that there are several methyl radicals that may be lost).

FIGURE 2.43 Mass spectrum of α-ionone.

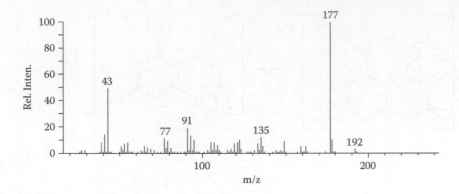

FIGURE 2.44   Mass spectrum of β-ionone.

$$+ C_4H_8 \qquad (2.76)$$

α-ionone   m/z 136

$$[M - C_2H_4]^+$$
$$m/z\ 164$$

β-ionone

$$-CH_3 \qquad (2.77)$$

or   or

m/z 177

   Examination of the two structures allows the results to be rationalized in terms of the relative stabilities of the respective molecular and fragment ions. The product ion from the retro Diels-Alder reaction from α-ionone is highly conjugated, while that from β-ionone is less stable, being cross-conjugated. The facile loss of a methyl radical from the latter compound can be rationalized as leading to a stable allylic carbonium ion or an aromatic-type ion in which the carbonyl oxygen has participated to form a ring. Under the concept of localized charge, the mechanism shown in Equation 2.78 appears reasonable to account for this ion.

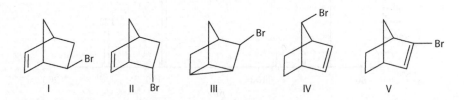

(2.78)

EXERCISE 2.8

The mass spectra of the isomeric bicyclic bromides C₇H₉Br are shown in Figures 2.45 to 2.49. Match each spectrum to one of the compounds.

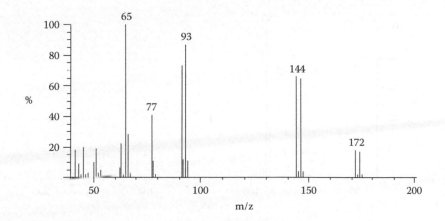

FIGURE 2.45 Exercise 2.8.

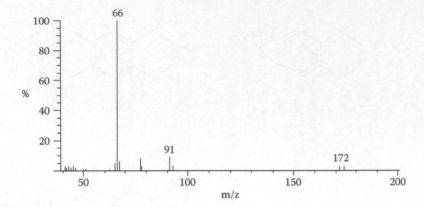

FIGURE 2.46   Exercise 2.8.

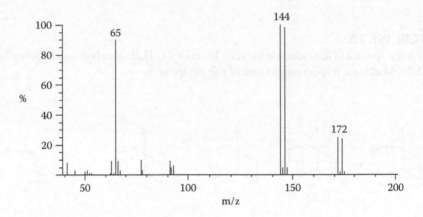

FIGURE 2.47   Exercise 2.8.

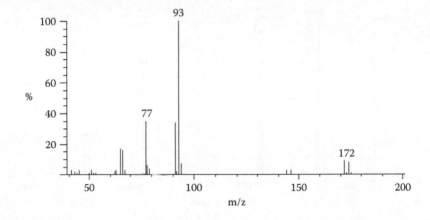

FIGURE 2.48   Exercise 2.8.

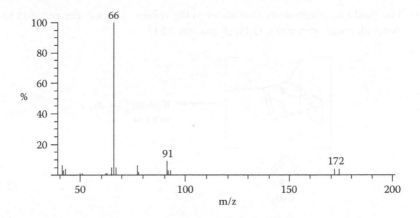

FIGURE 2.49   Exercise 2.8.

- Nortricyclyl bromide (III) is the only compound not containing a double bond. Thus, it should not yield an odd-electron (even-mass) ion corresponding to a retro Diels-Alder fragmentation.
- Figure 2.48 exhibits only odd-mass fragments and therefore belongs to this compound. The most intense peak in the spectrum arises by cleavage of the carbon-bromine bond to give the $C_7H_9^+$ ion at m/z 93 (Equation 2.79).

$$\longrightarrow\ C_7H_9^+\ +\ Br^\bullet \qquad (2.79)$$

III

- The retro Diels-Alder reaction from the two 5-bromonorbornenes (I and II) expels bromoethylene to give a $C_5H_6^+$ ion (Equation 2.80), such as is found in Figures 2.46 and 2.49.

$$\longrightarrow\ C_5H_6^+\ +\ C_2H_3Br \qquad (2.80)$$

$$m/z\ 66$$

I

- Cleavage of the carbon-bromine bond is in this case only of minor significance. It is impossible to differentiate by these spectra between the two isomers.

- The final two compounds also undergo the reverse Diels-Alder reaction to expel identical fragments, C_2H_4 (Equation 2.81)

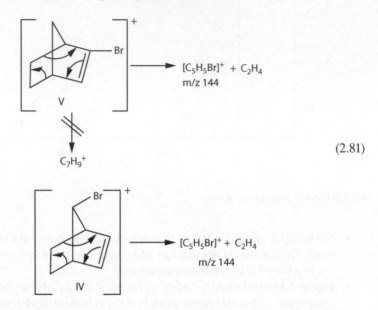

$$(2.81)$$

- In this case, however, the lower-mass fragment peaks offer evidence to identify the compounds.
- Figure 2.45 exhibits an intense peak at m/z 93, signifying loss of the bromine atom from the molecular ion, whereas Figure 2.47 has only a small peak at this mass.
- A choice between the two compounds is made on the basis that cleavage of the asp^2 carbon-bromine bond is not a particularly favorable process (the resultant vinyl carbonium ion is not a stable species).
- Compound IV, on the other hand, can easily lose its bromine atom (Equation 2.82).

$$(2.82)$$

Hydrogen Migrations

Hydrogen migrations appear with some frequency in the mass spectrometer, and considerable effort has been put forth to classify the scope of these reactions. In many cases, most notably in hydrocarbon and hydrocarbon-like materials, hydrogen migrations take place rapidly and randomly to make all the hydrogens in an ion essentially indistinguishable. However, in compounds containing functional groups that direct fragmentation, hydrogen migrations are usually specific and predictable. The observations that follow should be helpful in interpreting mass spectra when migrations are expected.

Hydrogen migrations occur most often in molecules containing rings or double bonds if simple cleavage fragmentations do not occur rapidly to lead to stable species. The McLafferty rearrangement may be considered a special case of this rule.

A five- or six-membered transition state is the favored process, although 1,2- and 1,3-migrations sometimes occur.

Hydrogen migrations usually result in a product ion that is more stable than the precursor or has a facile route to a stable ion.

Referring to the spectra of the cyclanols in Figures 2.28 to 2.30, let us examine the formation of the ion giving the intense peak at m/z 57. The common α-cleavage of alcohols, since the molecules are cyclic, leads to no loss of mass. Once this fragmentation occurs, however, in the case of the two larger ring compounds there is a convenient route to the highly conjugated ion a (Equation 2.83). Deuterium labeling has indeed established that the hydrogen indicated migrates across the ring and is lost in the fragmentation.

$$(2.83)$$

An unfavorable four-membered transition state is required to form the same ion from cyclobutanol; thus, a different fragmentation course is followed to yield the odd-electron ion of mass 44 (Equation 2.84).

$$(2.84)$$

The analogous fragmentation from cyclohexylamine and cyclohexanone leads to ions at m/z 56 and m/z 55, respectively.

EXPULSION OF STABLE NEUTRAL FRAGMENTS

The stability of the neutral fragment has generally been neglected in the preceding discussions, but such stability can often have considerable influence on the fragmentation course of a molecule. In every case, an odd-electron ion is formed that may be mistaken for a molecular ion; thus, anomalous results are sometimes obtained. Many acetoxy-containing compounds, for example, produce a peak 42 mass units below the molecular ion. According to our discussion so far, this would indicate a McLafferty rearrangement or retro Diels-Alder reaction to eliminate propylene (C_3H_6). In fact, this peak arises from the elimination of a molecule of ketene (C_2H_2O) by the transfer of a hydrogen atom from the acetyl group to the remainder of the molecule (Equation 2.85).

$$\left[R-O-\underset{\underset{H}{|}}{\overset{\overset{O}{\|}}{C}}-CH_2 \right]^{+} \longrightarrow [R-OH]^{+} + H_2C=CO \qquad (2.85)$$

$$M-42$$

Enol acetates have a particular propensity for this fragmentation, which may in these cases proceed through a six-membered transition state. Such a reaction could be placed in the category of McLafferty rearrangements (Equation 2.86).

$$\left[\begin{array}{c} RCH \overset{CH}{\diagdown} O \\ H \diagdown \underset{CH_2}{\diagup} \overset{C}{\diagdown} O \end{array} \right]^{+} \longrightarrow \left[\begin{array}{c} CHO \\ | \\ R-CH_2 \end{array} \right]^{+} + CH_2CO \qquad (2.86)$$

$$M-42$$

Even more striking is the spectrum of 3-bromo-3-phenylpropionic acid, which first loses a bromine radical, then ketene. The rearrangement fragmentation is supported by a metastable peak at m/z 76.8 ($107^2/149 = 76.84$). The stability of the benzylic product ion and the stability of the neutral ketene molecule provide the impetus for this rare rearrangement of a hydroxy group (Equation 2.87).

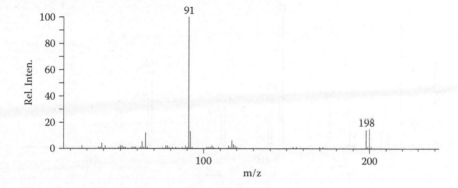

Likewise, the stability of ethylene provides driving force for its elimination from many compounds (including those that undergo the McLafferty and retro Diels-Aider reactions). One striking example is the expulsion of ethylene from the molecular ion of 1-phenyl-3-bromopropane (Figure 2.50). Deuterium-labeling studies have substantiated the mechanism shown for this fragmentation (Equation 2.88). We again see a rather rare migration, that is, of the bromine atom.

Without a further elaboration of the mechanisms (mainly because not much detail is known) or the scope of these reactions, let it suffice to say that fragmentations that

FIGURE 2.50 Mass spectrum of 1-phenyl-3-bromopropane.

expel neutral molecules such as CO, C_2H_4, CO_2, SO_2, and C_2H_2O are not uncommon. Some of these reactions may remind you of similar chemical reactions. The decarboxylation of β,γ-unsaturated acids, for example, is easily carried out simply by heating (Equation 2.89), and such compounds often exhibit an M–44 fragment ion.

$$R-CH=CH-CH_2-COOH \xrightarrow{\;e^-\; or\; heat\;} R-CH_2-CH=CH_2+CO_2 \quad (2.89)$$

Phthalic anhydride, when heated under vacuum in a flow tube to several hundred degrees, yields benzyne, which is detected by isolating its dimer (Equation 2.90). The mass spectral behavior of this compound is analogous, with an initial loss of CO_2 followed by loss of CO.

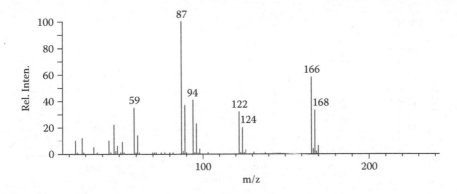

$$(2.90)$$

$[C_7H_4O]^+ \longrightarrow [C_6H_4]^+ + CO$
$+ CO_2 \qquad\qquad\quad m/z\ 76$

PROBLEMS 2.8–2.10
What are compounds that give the spectra in Figures 2.51 to 2.53 (Problems 2.8–2.10, respectively)?

The reactions outlined in this chapter form the basis for interpreting almost any mass spectrum. Various modifications or combinations of them, however, may tend to obscure the processes that occur in some complicated molecules. It is with practice that one will eventually be able to recognize these complications and actually

FIGURE 2.51   Problem 2.8.

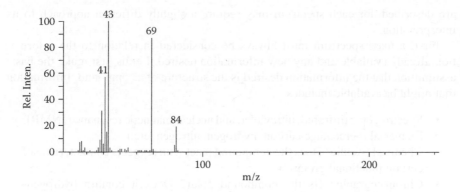

FIGURE 2.52   Problem 2.9.

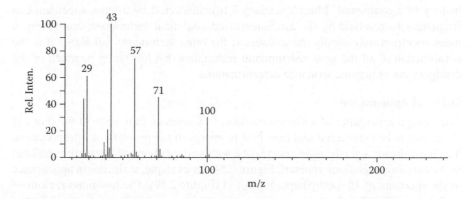

FIGURE 2.53   Problem 2.10.

use them in elucidating a structure. In addition, some rearrangement reactions not yet discussed may cause confusion. It is hoped that many of these rearrangements will soon be understood and placed into the basic framework of mass spectral "chemistry" just presented. To obtain a clearer understanding of how to use the mass spectrometer now requires practice, which the succeeding chapters are intended to give.

INTERPRETING EI MASS SPECTRA

Previous sections have attempted to provide a rational, basic understanding of the processes that occur in the mass spectrometer following electron bombardment. A stepwise development of the "chemistry" of fragmentation and a short introduction to the principles and instrumentation were deemed prerequisite to the full use of mass spectrometric data. We can now begin to examine mass spectra with the intention of elucidating the structures of unknown compounds. In this chapter, a general format for examining mass spectra is developed, or, more correctly, several formats are described, for each spectrum may require a slightly different approach to its interpretation.

First, a mass spectrum must always be considered in relation to the information already available and any new information desired. Let us first make the basic assumption that the information desired is the structure of a compound. Information that might be available includes

- Spectroscopic (infrared, ultraviolet, and nuclear magnetic resonance [NMR])
- Elemental (percentage carbon, hydrogen, nitrogen, etc.)
- Chemical (Acidic? Does the compound react with reagents characteristic of certain functional groups?)
- Chromatographic (Is the compound polar? Does it contain hydrogen-bonding groups?)

In addition, a considerable amount of information is available if one knows the history of a compound. Many questions left unanswered by a mass spectrum can frequently be resolved by the aforementioned analytical techniques; conversely, a mass spectrum may supply the answers if the other techniques fall short. It is the combination of all the new instrumental techniques that has taken so much of the drudgery out of organic structure determinations.

Spectral Appearance

The general appearance of a mass spectrum often gives an indication of the types of structures to be considered and how best to approach the problem of interpretation. You have already seen the mass spectra of a wide variety of organic structures and can make certain correlations yourself. Figure 2.54, for example, is similar in appearance to the spectrum of 15-methylheptadecan-1-ol (Figure 2.19). The low-mass regions of both spectra contain the more intense peaks, and the molecular ions are of low or negligible intensity. The spectra contain groups of peaks separated by about 14 mass units, the mass of a methylene unit (—CH$_2$—). These are typical "hydrocarbon" spectra, obtained from compounds having a large number of bonds with nearly equal strengths. No specific fragmentation can lead to a particularly stable fragment ion.

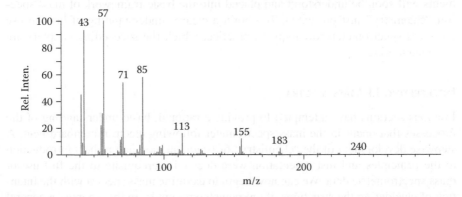

FIGURE 2.54　Mass spectrum of 2,6,10-trimethyltetradecane.

However, as stated previously, the spectra of hydrocarbons can sometimes yield information concerning the nature of the carbon skeleton. There is a slight preference for cleavage at the branched position of hydrocarbons, which leads to relatively intense peaks corresponding to the secondary or tertiary ion products. The spectrum in Figure 2.18 was chosen as a clear indication of this favored type of cleavage. Figure 2.54 also contains relatively intense peaks, at m/z 155 and 183, that indicate branching of the molecule. These peaks are conspicuous by their greater intensity compared with their homologous neighbors.

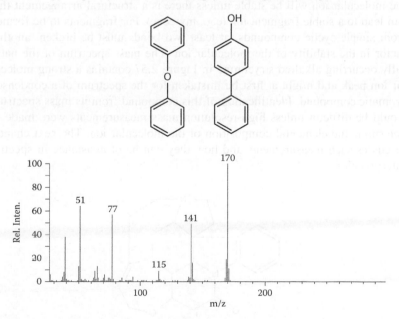

The spectra of aromatic hydrocarbons have a much different appearance. The ease with which these molecules with their π-electron systems accommodate the loss of an electron results in very intense molecular ion peaks. The only significant fragmentation is that which cleaves an alkyl chain attached to the aromatic ring. In these spectra, the molecular weight derived from the mass of the molecular ion may be sufficient evidence to confirm a structure. The nominal mass (170) of the molecular ion in Figure 2.55, for example, can correspond to only one composition that conforms to an aromatic-type molecule, that is, $C_{12}H_{10}O$. Although we cannot differentiate among the possible isomers using the mass spectrum alone, it should be a simple problem for NMR or infrared spectroscopy. Diphenyl ether is the compound that gives this spectrum.

FIGURE 2.55 Mass spectrum of diphenyl ether.

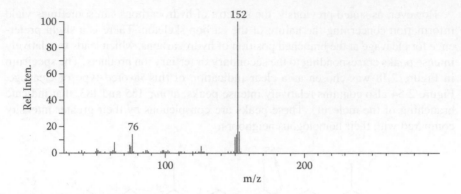

FIGURE 2.56   Problem 2.11.

PROBLEM 2.11

What is the structure of the compound that gives the spectrum in Figure 2.56?

The more typical organic compound gives a spectrum between these two extremes in appearance. Although it is impossible to classify structural types further with respect to the general appearance of their mass spectra, a few comments might be helpful.

Conjugated olefins and cyclic compounds tend to have fairly intense molecular ion peaks. Cleavage of the vinylic or double bonds of olefins is difficult, so the molecular ion will be stable unless there is a structural arrangement that can lead to a stable fragment ion (e.g., myrcene). For fragments to be formed from simple cyclic compounds, at least two bonds must be broken, another factor in the stability of the molecular ion. The mass spectrum of the naturally occurring alkaloid strychnine in Figure 2.57 contains a strong molecular ion peak and might at first be mistaken for the spectrum of a condensed aromatic compound. Identification of this compound from its mass spectrum would be difficult unless high-resolution mass measurements were made to determine the elemental composition of the molecular ion. The next chapter discusses such measurements and how they can be of assistance in spectral interpretation.

Strychnine

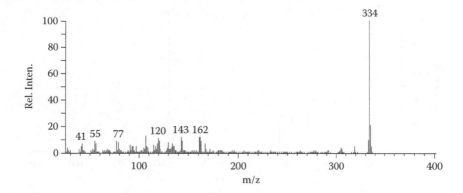

FIGURE 2.57   Mass spectrum of strychnine.

Heteroatoms in a molecular ion tend to promote fragmentation. Cleavage of
the carbon-heteroatom bond and of the bond α to the heteroatom is common.
n-Decanol, for example, has a relative molecular ion intensity nearly 1,000
times less than that of *n*-decane, and the expulsion of HCN from pyridine
(Figure 2.58b) is considerably more facile than loss of C_2H_2 from benzene
(Figure 2.58a), depicted by Equation 2.91.

$$\text{(2.91)}$$

Isotopes

The use of isotope peaks to calculate elemental compositions of molecular ions was
discussed previously. This technique may also be applied to fragment ions but is in
general only of secondary importance; the $(P + 1)$ and $(P + 2)$ peaks usually arise
not only from the isotope contribution but also from a significant contribution of ions
having altogether different compositions. The recognition of isotope peak patterns
arising from the presence of certain elements can, however, be extremely valuable.
The isotopic distribution of chlorine (76% ^{35}Cl and 24% ^{37}Cl) is easily recognized in
a spectrum, as that of bromine is (50% ^{79}Br and 50% ^{81}Br). Silicon and sulfur also
provide characteristic $(P + 2)$ isotope peaks.

Although we will not particularly want to use it for calculating the complete elemental compositions of ions, the rules presented in Equations 2.1 and 2.2 are of some
general value. That is, *each atom in an ion will contribute an amount to the intensity*

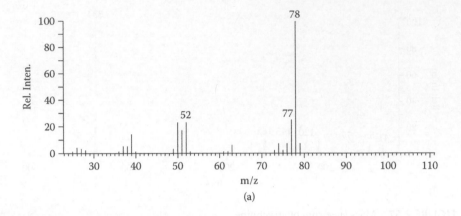

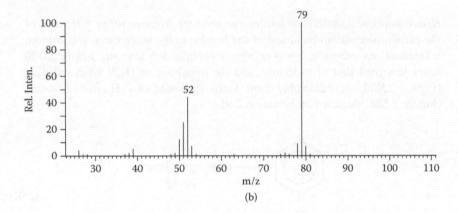

FIGURE 2.58   Mass spectra of (a) benzene and (b) pyridine.

of the isotope peaks that is equal to the relative abundances of the isotopes of that element. Thus, tetramethylsilane $Si(CH_3)_4$ exhibits at the molecular ion $(P + 1)$ and $(P + 2)$ peak intensities of 9.1% and 3.1% $(P = 100\%)$, respectively, while the *bis*-(trimethylsilyl) derivative of ethylene glycol has peak intensities of 100%, 18.2%, and 7.0% at the molecular ion.

Once a pattern is recognized and related to a specific heteroatom or combination of heteroatoms, subtraction of the known mass often leaves the mass of a common fragment as listed in Table 2.7. To go further than this is usually not warranted since only by scanning the peaks of interest several times to average electronic and ion statistic effects or by use of a double-collector system (to measure the relative intensities of a P and a $(P + 1)$ or $(P + 2)$ peak simultaneously) can intensities be measured with the accuracy required to calculate complete compositions.

$$(CH_3)_3SiO \qquad OSi(CH_3)_3$$
$$| \qquad\qquad |$$
$$H_2C \longrightarrow CH_2$$

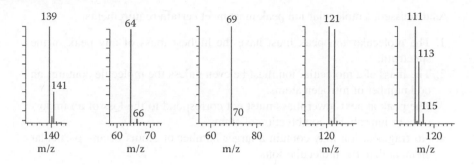

FIGURE 2.59   Isotope clusters for Exercise 2.9 and Problem 2.12.

206	100%	P	
207	18.2%	$P + 1$	($C_8 = 8.8$ and $Si_2 = 9.4$)
208	7.0%	$P + 2$	

EXERCISE 2.9

What are the compositions of the ions that give the isotope peak patterns in Figure 2.59? They are not necessarily molecular ions.

- The $(P + 2)$ peak of example (a) is close to one-third the intensity of the P peak.
- A chlorine atom is therefore indicated.
- Subtraction of the 35 Da from the ion mass 139 Da leaves 104 Da.
- This is the mass of a benzoyl ion minus the one hydrogen that was replaced by the chlorine, C_7H_4O.

m/z 139

PROBLEM 2.12

Solve the remaining examples in Figure 2.59.

Molecular Ions

Identification of the molecular weight is the most crucial step in the proper interpretation of a mass spectrum. A wrong assignment here will lead to many fruitless attempts to correlate the fragment ions with a structure; a proposed structure cannot be correct if the molecular weight assignment is in error. This does not mean that a molecular ion peak must necessarily be observed. We now discuss how to recognize molecular ions and how to determine a molecular weight when no peak is seen at this mass.

As discussed, a molecular ion peak must meet certain requirements:

1. The molecular ion peak must have the highest mass of any peak in the spectrum.
2. The mass of a molecular ion must be even unless the molecule contains an odd number of nitrogen atoms.
3. The peak at next lower mass must not correspond to the loss of an impossible or improbable combination of atoms.
4. No fragment ion may contain a larger number of atoms of any particular element than the molecular ion.

These rules will not necessarily identify the molecular ion peak; rather, they are meant more for the exclusion of incorrect ions than for actual identification of the correction.

A molecular weight can often be determined or substantiated by examining the fragmentation behavior of a compound. Two fragment peaks may give direct evidence of the molecular weight if they correspond to ions containing different parts of the original molecule. The simplest case is that of the molecule's readily breaking into two pieces, both of which form ionic species. The methyl ester of phenylalanine (Figure 2.60) exhibits a weak molecular ion, but three important fragment ions are formed by cleavage α to the nitrogen function. Two of them, at *m/z* 88 and 91, lead to the correct molecular weight by combination of their masses (Equation 2.92).

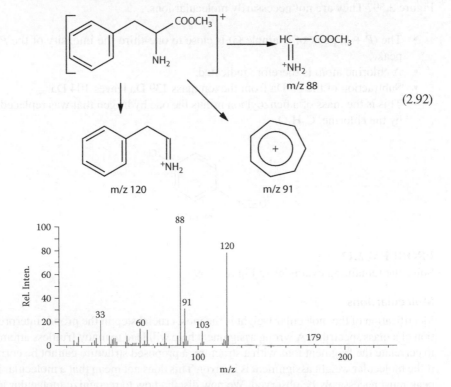

$$(2.92)$$

FIGURE 2.60   Mass spectrum of phenylalanine methyl ester.

Branched alcohols seldom exhibit peaks at the masses corresponding to their molecular weights. In addition to the elimination of water from the molecular ion, α-cleavages on either side of the hydroxy group are favored fragmentations. Nonetheless, molecular weights and structures can usually be determined. The examples in Problem 2.13 illustrate this point.

PROBLEM 2.13

The spectra of the three alcohols that follow are given in Figures 2.61–2.63. Match the spectra with the structures and provide rationalization for your choices.

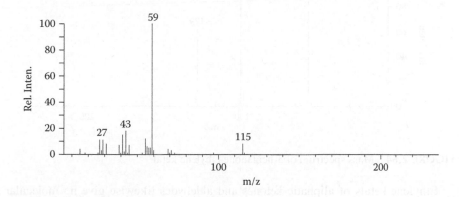

FIGURE 2.61   Problem 2.13.

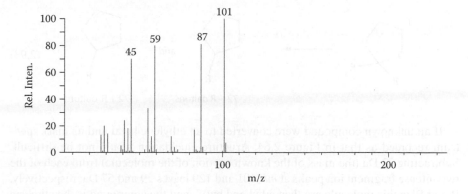

FIGURE 2.62   Problem 2.13.

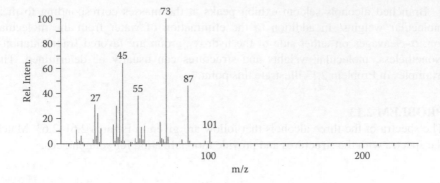

FIGURE 2.63   Problem 2.13.

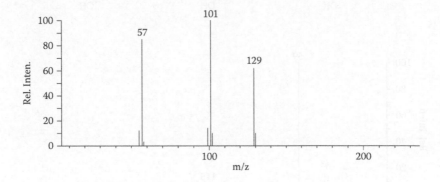

FIGURE 2.64   Mass spectrum of 3-heptanone ethylene ketal.

Ethylene ketals of aliphatic ketones and aldehydes likewise give no molecular ion peaks. The characteristic α-cleavage, which results in the highly stable oxonium ions, however, can be used to identify the molecular weights and structures of many such compounds (Equation 2.93).

(2.93)

72 + R daltons   72 + R' daltons

If an unknown compound were converted to an ethylene ketal and its mass spectrum recorded as that in Figure 2.64, structure elucidation would not be difficult. Subtracting 72 Da (the mass of the known portion of the molecule) from each of the two intense fragment ion peaks at m/z 101 and 129 leaves 29 and 57 Da, respectively. R and R' in this molecule are then ethyl and butyl, and the compound is the ethylene ketal of 3-heptanone (Equation 2.94).

$$(2.94)$$

C₂H₅
m/z 101   m/z 129

Fragment Ions

Determination of the molecular weight is an important first step in the interpretation of a mass spectrum and may in fact give all the information necessary for the elucidation of a structure. However, the greatest value of mass spectrometry lies in the quantity of data that can be derived from the degradation products formed after electron bombardment. Just as chemical degradation products have been used in the past (and are still used in many problems) to obtain information about small portions of a complex molecule, fragment ions can often be used to completely reconstruct a molecular structure. There are several ways to examine the fragment ion peaks, all of which may be useful in specific cases. Many people like to start at the molecular ion and proceed to each lower mass peak to see what small units are easily lost (hydrogen, methyl, ethyl and other radicals, water, HX, etc.) and the mechanism of the loss (simple cleavage or multicenter).

In Figure 2.65, we see the mass spectrum of a compound with a molecular weight of 122. The very intense peak at m/z 91 is indicative of the tropylium ion and has been formed by the loss of 31 mass units from the molecular ion. A glance at Tables 2.10 and 2.11 reveals that only two fragments can be lost to account for this mass difference: NH_2CH_3 and OCH_3. The first would require a molecular composition of $C_7H_{10}N_2$ (and C_6H_5N, not C_7H_7, for m/z 91). The more likely candidate is methyl benzyl ether. The isomeric anisoles can also be excluded on the basis that they have no labile carbon-oxygen bond (Equation 2.95).

$$(2.95)$$

$C_7H_7^+$, m/z 91

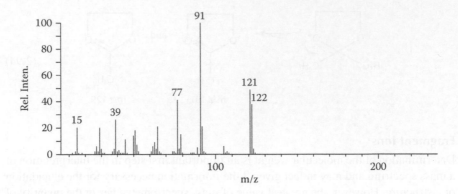

FIGURE 2.65 Mass spectrum of methyl benzyl ether.

TABLE 2.10
Neutral Fragments Expelled by Simple Cleavage

Fragment	Nominal Mass	Fragment	Nominal Mass
H	1	CH_2SH	47
CH_3	15	CH_2Cl	49
NH_2	16	CHF_2	51
OH	17	CH_2CH_2CN	54
F	19	C_4H_7; C_3H_3O	55
CN	26	C_4H_9; C_3H_5O	57
C_2H_3	27	C_3H_8N	58
C_2H_5; CHO	29	C_3H_7O, $COOCH_3$; CH_2COOH	59
CH_3O; CH_2OH	30	C_2H_5S	61
CH_3O; CH_2NH_2	31	C_5H_5	65
SH; CH_2F; $(H_2O + CH_3)$	33	CF_3	69
Cl	35	C_4H_7O	71
CH_2CN	40	C_6H_5	77
C_3H_5	41	Br; C_6H_7	79
C_3H_7; CH_3CO	43	C_6H_9; C_5H_5O	81
C_2H_5NH	44	C_7H_7	91
C_2H_5O; COOH	45	C_7H_5O; C_8H_9	105
NO_2	46	I	127

Table 2.10 contains small mass fragments that are commonly lost by simple cleavage and are indicative of specific structural features. In most cases, an intense even-electron (odd-mass, unless nitrogen is present) fragment ion peak indicates that the unit lost is indeed present in the molecule. Rearrangements, unless they yield exceptionally stable ions, are generally slower reactions and lead to weaker peaks.

Note that several of the mass differences can be accounted for by more than one possible structure, or even composition. An M–31 ion may be indicative of a methoxy substituent (as in the example of methyl benzyl ether), but the same atoms may be lost if there is a primary alcohol function in a molecule (Equations 2.96 and 2.97).

$$R \overset{\curvearrowright}{\underset{}{\text{—}}} \overset{+}{O}CH_3 \quad\longrightarrow\quad R^+ + \overset{\bullet}{}OCH_3 \tag{2.96}$$

$$R \overset{\curvearrowright}{\underset{}{\text{—}}} CH_2 \text{—} {}^+OH \quad\longrightarrow\quad R^+ + \overset{\bullet}{}CH_2OH \tag{2.97}$$

Simple cleavage with the loss of 43 mass units may be indicative of an acetyl or a propyl unit in the molecule. Isotope peak intensities may help to identify the correct composition, but of greater value in this respect is high-resolution mass measurement. From the accurate masses of the molecular and fragment ions, the elemental composition of a neutral fragment can be determined. More about this important technique is discussed in Chapter 3.

EXERCISE 2.10
What is the compound that gives the spectrum in Figure 2.66?

- As in the preceding spectrum, the intense molecular ion and the scarcity of fragment peaks are indicative of stabilization by an aromatic ring or highly conjugated olefinic system.
- This compound is an isobar (has the same molecular weight) of methyl benzyl ether ($M = 122$), but the molecular ion fragments by loss of 17 mass units rather than 31.
- It is likely, then, that a hydroxy function is present in an environment that facilitates simple cleavage.
- Looking to Table 2.10, we see two possible compositions (and several structures) for the even-electron ion of mass 105: C_8H_9 and $C_7H_5O^+$.

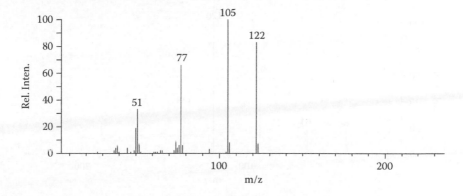

FIGURE 2.66   Exercise 2.10.

- The choice becomes clear when the next-lower fragment peak is considered, for only the benzoyl ion loses 28 mass units (as CO) readily (Equation 2.98). The compound is therefore benzoic acid.

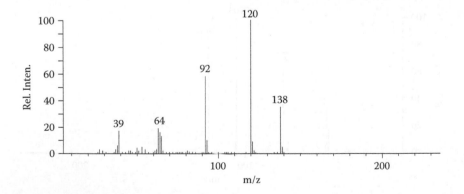

$$ \text{(2.98)} $$

At first glance, the addition of an *ortho* hydroxyl group to benzoic acid would seem to have little effect on the fragmentation behavior of the molecule (see Figure 2.67).

However, the intense fragment peak immediately below the molecular ion has an even mass and corresponds not to a simple cleavage product but to the multicenter elimination of water. Such an *ortho* effect is not uncommon in mass spectrometry and can be rationalized by a six-membered transition state. The subsequent loss of CO may occur by expulsion of the group from the ring or from the new ketene function (Equation 2.99). *Meta* and *para* isomers of this compound fragment in a manner parallel to that of benzoic acid: loss of a hydroxy radical followed by loss of CO (Equation 2.100).

$$ \text{(2.99)} $$

FIGURE 2.67   Mass spectrum of salicylic acid.

$$\left[\;\underset{\overset{|}{OH}}{\overset{\overset{\displaystyle CO-OH}{|}}{\bigcirc}}\;\right]^{+} \longrightarrow \underset{\overset{|}{OH}}{\overset{\overset{\displaystyle CO^{+}}{|}}{\bigcirc}} \xrightarrow{\;-CO\;} C_6H_5O^{+} \qquad (2.100)$$

m/z 138   m/z 121   m/z 93

Yet another example of an *ortho* effect is the series of isomeric sulfur-containing compounds whose spectra are shown in Figure 2.68.

The three spectra are significantly different and are rationalized as follows:

- The *ortho* isomer, due to the steric and electronic interaction of the neighboring groups, has a relatively strong tendency to lose either the methyl radical or a molecule of CO_2.
- The *meta* isomer, on the other hand, exhibits little fragmentation.
- The *para* isomer is the CO_2 least sterically hindered, but the S-methyl group can now stabilize an M–OH ion by resonance. Thus, the peak at *m/z* 151 is relatively intense.

$$+SCH_3$$

m/z 151

The fragments expelled by multicenter fragmentations form a second list, which is given in Table 2.11. Some of the masses in this list are the same as in Table 2.10 but correspond to different compositions. One must always be cognizant of this possible ambiguity, but again, high-resolution mass measurements could eliminate from consideration all but the correct composition.

PROBLEM 2.14

The spectra in Figures 2.69 and 2.70 are of compounds related to salicylic acid. What are their structures?

A second fruitful approach to the interpretation of a mass spectrum is to examine the low-mass fragment peaks (below, say, *m/z* 150) for characteristic ions.

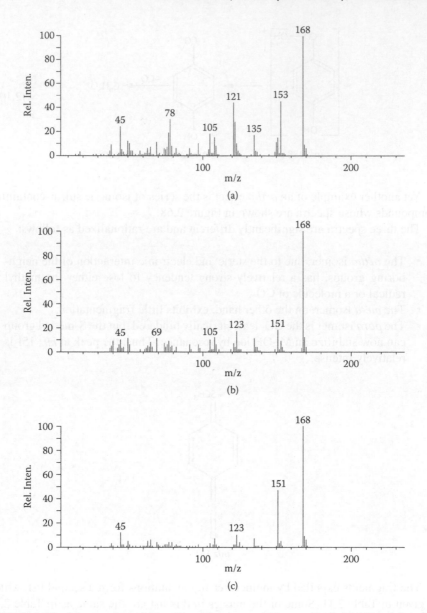

FIGURE 2.68 Mass spectra of (a) *ortho*-, (b) *meta*-, and (c) *para*-(methylthio)benzoic acids.

Some of these have already been discussed and are summarized along with others in Table 2.12. The substituted tropylium ions are often indicators of the correspondingly substituted benzylic group in a molecule. Similar ring-enlarged fragments from pyrroles and furans have also been mentioned. Another is the $C_5H_9O^+$ fragment at *m/z* 85 usually found in the mass spectra of tetrahydropyranyl ethers—often used in organic chemistry as base-stable protecting groups for hydroxyl functions.

TABLE 2.11
Neutral Fragments Expelled by Multicenter Fragmentations

Fragment	Nominal Mass	Fragment	Nominal Mass
H_2	2	C_2H_7N	45
NH_3	17	C_2H_6O, $(H_2O + C_2H_4)$	46
H_2O	18	CH_4S	48
HF	20	C_4H_6	54
HCN	27	C_4H_8; C_3H_4O	56
CO; C_2H_4	28	C_3H_6O	58
CH_2O	30	C_3H_9N	59
CH_5N	31	C_3H_8O; C_2H_4O	60
CH_4O	32	C_2H_6S	62
H_2S	34	$C_3H_6O_2$	74
HCl	36	C_6H_4	76
C_3H_6; C_2H_2O	42	C_6H_6	78
CO_2	44	HBr	80

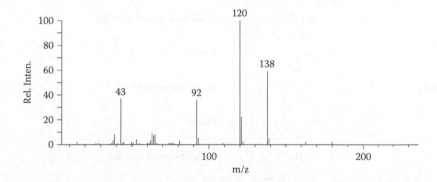

FIGURE 2.69   Problem 2.14.

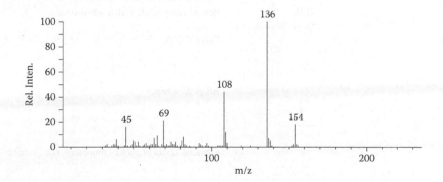

FIGURE 2.70   Problem 2.14.

TABLE 2.12
Characteristic Low-Mass Fragment Ions

Nominal Mass	Ion	Origin
43	$C_3H_7^+$	Alkyl groups
	$C_2H_3O^+$	Acetyl group
58	$[H_2C\!=\!\overset{\overset{\displaystyle OH}{\mid}}{C}\!-\!CH_3]^+$	Ketones
59	$^+COOCH_3$	Methyl esters
61	$CH_3\overset{\overset{\displaystyle OH}{\mid}}{C}\!=\!\overset{+}{O}H$	Esters of high molecular weight alcohols
70	(pyrrolinium ring, $\overset{+}{N}H$)	Pyrrolidines
74	$[H_2CO\!=\!\overset{\overset{\displaystyle OH}{\mid}}{C}\!-\!CH_3]^{+\bullet}$	Methyl esters with a γ-hydrogen
78	$C_5H_4N^+$	Pyridines and alkyl pyrroles
80	(pyrrole ring, $\overset{+}{N}H$)	Pyrroles
79(81)	Br^+	Bromo compounds
81	(furan ring, $\overset{+}{O}$)	Furans
83	$C_6H_{11}^+$	Cyclohexanes or hexenes
85	(tetrahydropyran ring, $\overset{+}{O}$)	Tetrahydropyranyl ethers
91	$C_7H_7^+$	Aromatic hydrocarbons with side chains
92	$C_7H_8^+$	Benzyl compounds with a γ-hydrogen
95	(furyl–C≡O^+)	Furyl–CO–X
97	(thiophene ring, $\overset{+}{S}$)	Alkyl thiophenes
99	(ethylene ketal ring, O, $\overset{+}{O}$)	Ethylene ketals of cyclic compounds (i.e., steroids)

TABLE 2.12 (*Continued*)
Characteristic Low-Mass Fragment Ions

Nominal Mass	Ion	Origin
104	$C_8H_8^+$	Alkyl aromatics
105	$C_7H_5O^+$	Benzoyl compounds
	$C_8H_9^+$	Aromatic hydrocarbons
106	$H_2C=\!\!\langle\!\!\rangle\!\!=NH_2^+$	Amino benzyl
107	$C_7H_7O^+$	Phenolic hydrocarbons
117	$C_9H_9^+$	Styrenes
127	I^+	Iodocompounds
130	(indole structure)	Indoles
131	(cinnamoyl structure)	Cinnamates
149	(phthalic structure)	Dialkyl phthalates (by rearrangement)

Any ion in this table may, of course, be substituted in such a way as to make it nearly unrecognizable. Only with practice can one make the proper assignments. In addition, other spectral data can play a useful role in this respect. For example, if the NMR spectrum of an unknown contains a peak characteristic of a methoxy group on an aromatic ring, one would look for a tropylium ion at *m/z* 121, not at 91.

Without high-resolution mass measurements, a question often arises concerning the composition of an ion, whether it is a hydrocarbon ion or a more unsaturated oxygen-containing ion. Isotope peak ratios or subsequent fragmentations may provide clues to the solution of this question. In aromatic systems, for example, the tropylium and benzoyl homolog series correspond in mass, but the latter ions generally fragment further by loss of a molecule of carbon monoxide, 28 Da (Equation 2.101). Tropylium ions lose an acetylene fragment to only a slight extent (Equation 2.102). The spectrum of acetophenone in Figure 2.71 can be compared with that of methylethylbenzene (Figure 2.21).

$$\text{(reaction scheme)} \quad \xrightarrow{-CO} C_6H_4X^+ \qquad (2.101)$$

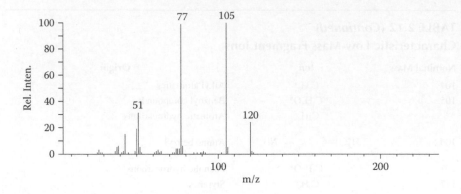

FIGURE 2.71   Mass spectrum of acetophenone.

(2.102)

The propyl ($C_3H_7^+$) fragment usually fragments further in a manner typical of most saturated hydrocarbon ions by loss of H_2. This leads to intense peaks at m/z 41 and 39.

The acetyl ($C_2H_3O^+$) fragment, on the other hand, is accompanied by only minor peaks at the lower masses.

PROBLEMS 2.15 AND 2.16

Identify the compounds that give the spectra in Figures 2.72 and 2.73 (Problems 2.15 and 2.16, respectively).

EXERCISE 2.11

Identify the compound that gives the spectrum in Figure 2.74.

- The molecular ion peak *(m/z 231)* in this spectrum offers several pieces of structural information.
  - Its intensity is indicative of stabilization by conjugation or fused rings.
  - The odd mass tells us there is at least one nitrogen atom present.
  - The (*P* + 2) isotope peak is clearly due to the presence of one chlorine atom. Do not be confused by the M–H ion *(m/z 230)* and its isotope peak *(m/z 232)*.

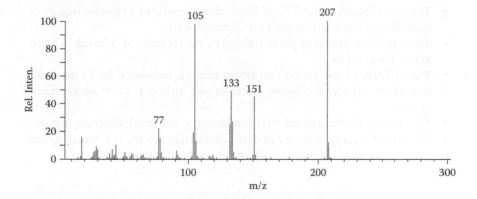

FIGURE 2.72 Problem 2.15.

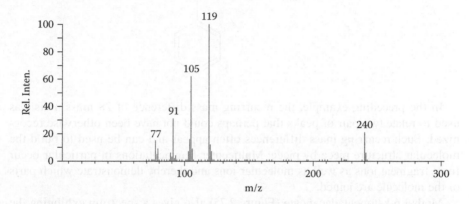

FIGURE 2.73 Problem 2.16.

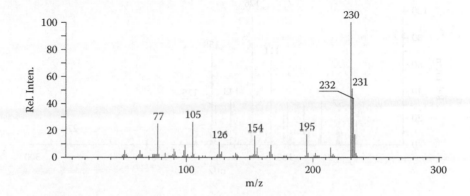

FIGURE 2.74 Exercise 2.11.

- The pair of peaks at *m/z* 77 and 105 is characteristic of a benzoyl moiety (a methyltropylium ion does not lose 28 mass units).
- There is a similar pair of peaks (related by the presence of chlorine in both) at *m/z* 126 and 154.
- The difference between the two sets of peaks is accounted for by the addition of an NH and a chlorine (subtract one hydrogen when adding each group).
- The entire molecule is therefore characterized as aminochlorobenzophenone.
- Based on this spectrum, we can state conclusively that the two functions are on the same ring, but their relative positions remain unknown.

In the preceding example, the recurring mass difference of 28 mass units was used to relate two pair of peaks that perhaps could not have been otherwise recognized. Such recurring mass differences often appear and can be used to build the molecular structure piece by piece. Multicenter fragmentations in particular occur from fragment ions as well as molecular ions and thereby demonstrate which parts of the molecule are joined.

Methyl 6-ketopentadecanoate (Figure 2.75) also gives a spectrum exhibiting the recurrence of methanol elimination. The molecular ion, an odd-electron ion at *m/z*

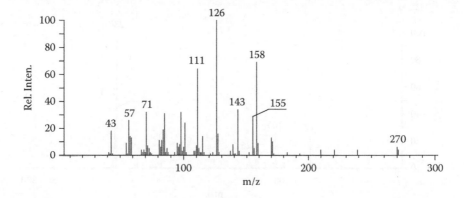

FIGURE 2.75   Mass spectrum of methyl 6-ketopentadecanoate.

158, and an even-electron ion of mass 143 are related to one another by this elimination (Equation 2.103).

$$CH_3(CH_2)_5CHCH_2CH_2CCH_2CH_2CH_2CH_2COOCH_3$$

m/z 270

$$[H_2C=CCH_2CH_2CH_2CH_2COOCH_3]^+$$

m/z 158

$$CCH_2CH_2CH_2CH_2COOCH_3 \quad (2.103)$$

m/z 143

$$M^+ (m/z\ 270) \longrightarrow 238^+ + CH_3OH$$
$$158 \longrightarrow 126^+ + CH_3OH$$
$$143 \longrightarrow 111^+ + CH_3OH$$

The odd-electron ion is formed by a McLafferty rearrangement on the alkyl side of the keto group, while the ion of mass 143 arises by simple cleavage α to the carbonyl. The common elimination of methanol from these ions demonstrates that each contains the methyl ester function.

Another approach to mass spectral interpretation is to compare the spectrum of an unknown with those of compounds believed to be related to the unknown. This method is particularly applicable in natural product studies, in which complete fragmentation patterns are difficult to rationalize. The large and complex molecules often have a tendency to rearrange either before or after electron. The spectra are nonetheless reproducible and can be used for comparative purposes.

A study of the iboga alkaloids is the classic example of this type of approach to mass spectrometry.

Without a detailed discussion of the mass spectra, two series of ions can be identified by comparison of the spectra.

- One series contains fragments of the indole part of the molecule, and the other contains fragments of the more saturated isoquinuclidine system. The ions containing the indole nucleus are shifted in mass when the substitution on the aromatic ring is varied (195 → 225 → 255; 280 → 310 → 340).
- The other ion series remains constant in each spectrum (122, 136, 149 Da). The spectra of the unsubstituted and monosubstituted compounds (Figures 2.76a, 2.76b) were used to elucidate the structure of the third compound, the dimethoxy derivative (Figure 2.76c).

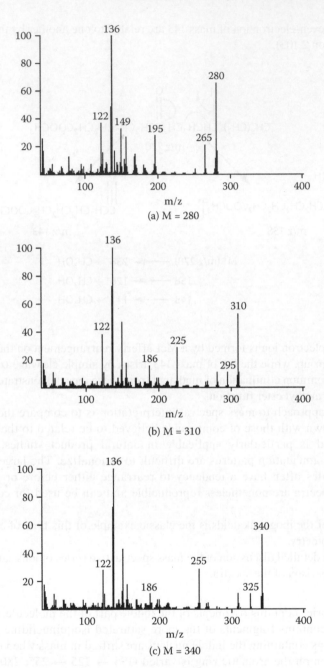

FIGURE 2.76 Mass spectra of substituted iboga alkaloids.

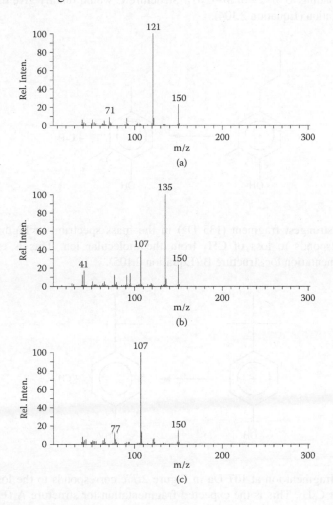

$X = Y = H$
$X = OCH_3; Y = H$
$X = Y = OCH_3$

EXERCISE 2.12

Assign the mass spectra shown in Figure 2.77a–2.77c to the phenols A, B, and C shown in the following:

FIGURE 2.77 Exercise 2.12.

- All three mass spectra exhibit the expected molecular ion at 150 Da.
- The spectrum in Figure 2.77a exhibits one major fragment at 121 Da, corresponding to $M-29$ or $M-C_2H_5$. Structure C would readily give this fragmentation (Equation 2.104).

$$\text{(2.104)}$$

- The strongest fragment (135 Da) in the mass spectrum in Figure 2.77b corresponds to loss of CH_3 from the molecular ion. This is expected fragmentation for structure B (Equation 2.105).

$$\text{(2.105)}$$

- The fragment ion at 107 Da in Figure 2.77c corresponds to the loss of 43 Da or C_3H_7. This is the expected fragmentation for structure A (Equation 2.106).

$$(2.106)$$

EXERCISE 2.13

Assign the mass spectra shown in Figures 2.78a–2.78c to the alcohols A, B, and C shown in the following:

A   B   C

- These alcohols all have the same molecular formulas as the phenols just reviewed, and molecular ions are seen at 150 Da in all cases.
- The spectrum in Figure 2.78b exhibits a single strong fragment at 107 Da (M–43 or M–C$_3$H$_7$). Structure C contains a C$_3$H$_7$ unit that is benzylic and α to a hydroxyl function, both providing driving force for this fragmentation (Equation 2.107).

107

$$(2.107)$$

- The spectrum in Figure 2.78c exhibits a single strong fragment at 119 Da (M–31 or M–CH$_2$OH). Structure B contains a primary alcohol function that is easily lost, again giving a benzylic fragment (Equation 2.108).

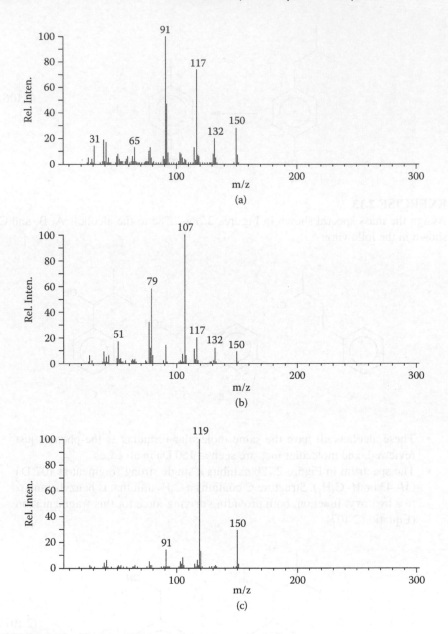

FIGURE 2.78   Exercise 2.13.

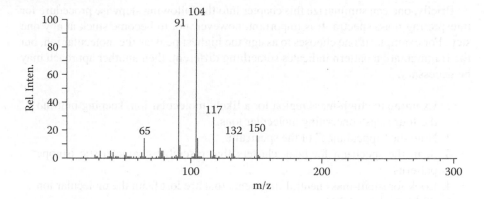

FIGURE 2.78D Problem 2.17.

$$(2.108)$$

- The spectrum in Figure 2.78a is more complex, indicating a structure that may fragment in several ways. Structure A provides the fragmentation paths shown in Equation 2.109.

$$(2.109)$$

PROBLEM 2.17
What is the compound that gives the spectrum in Figure 2.78d?

Briefly, one can summarize this chapter into the following stepwise procedure for interpreting mass spectra. It is important, however, not to become stuck at any one step. For example, if one chooses to assign the highest peak as the molecular ion, but the fragmentation pattern indicates something different, then another approach may be necessary.

1. Examine the high-mass region for a likely molecular ion, keeping in mind the four rules concerning molecular ions.
2. Note the "appearance" of the spectrum.
3. Scan the spectrum for peak clusters that resemble characteristic isotope patterns.
4. Look for small-mass neutral fragments that are lost from the molecular ion (Tables 2.10 and 2.11).
5. Look for characteristic low-mass fragment ions (Table 2.12).
6. Compare the spectrum with those of compounds believed to be related to the unknown.

PROBLEMS 2.18–2.34
Provide structures for the compounds that give the mass spectra in Figures 2.79 to 2.95 (Problems 2.18–2.34, respectively).

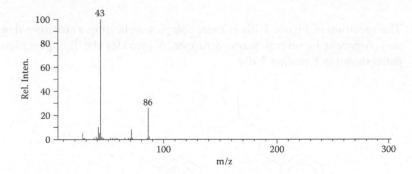

FIGURE 2.79 Problem 2.18.

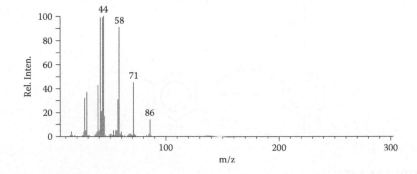

FIGURE 2.80 Problem 2.19.

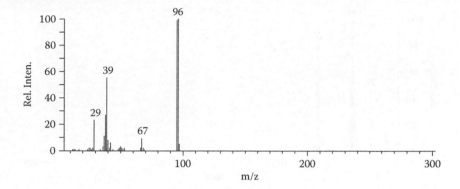

FIGURE 2.81   Problem 2.20.

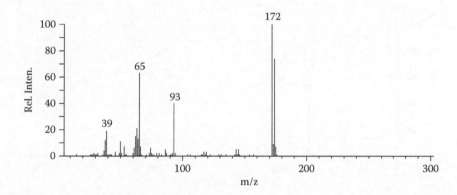

FIGURE 2.82   Problem 2.21.

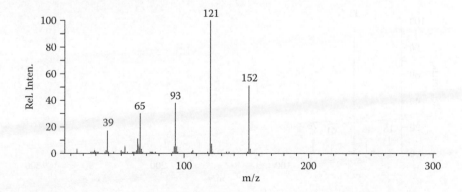

FIGURE 2.83   Problem 2.22.

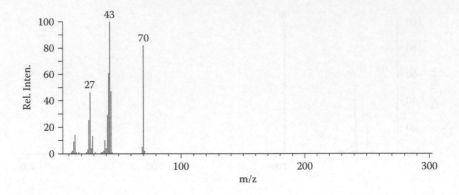

FIGURE 2.84   Problem 2.23.

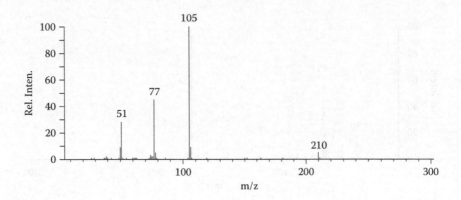

FIGURE 2.85   Problem 2.24.

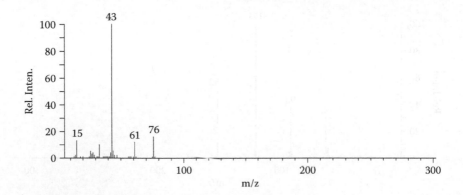

FIGURE 2.86   Problem 2.25.

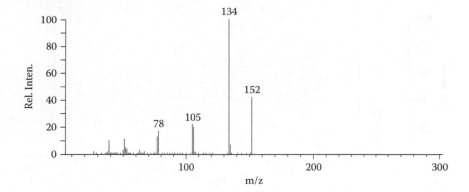

FIGURE 2.87 Problem 2.26.

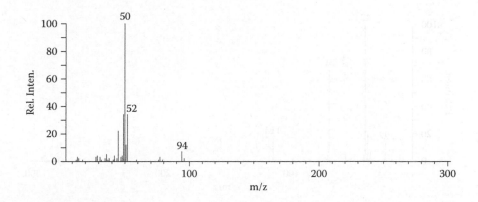

FIGURE 2.88 Problem 2.27.

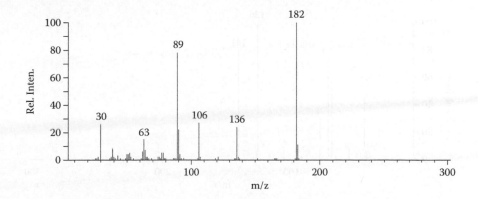

FIGURE 2.89 Problem 2.28.

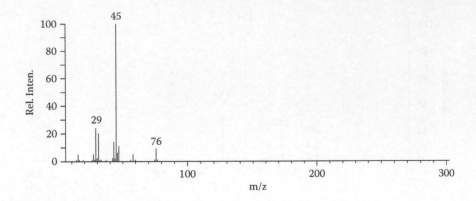

FIGURE 2.90 Problem 2.29.

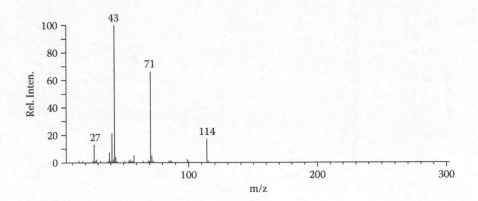

FIGURE 2.91 Problem 2.30.

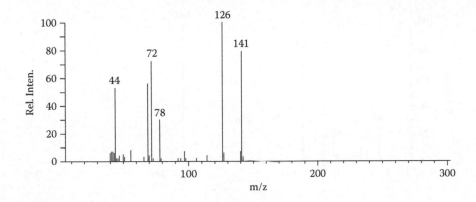

FIGURE 2.92 Problem 2.31.

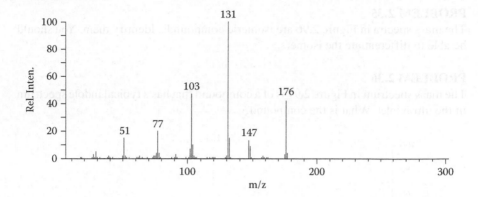

FIGURE 2.93   Problem 2.32.

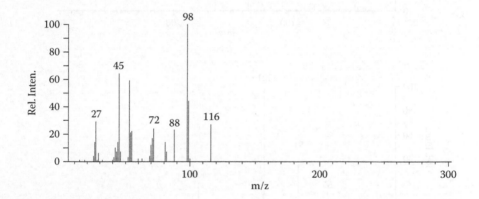

FIGURE 2.94   Problem 2.33.

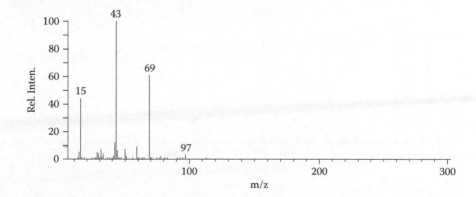

FIGURE 2.95   Problem 2.34.

PROBLEM 2.35

The mass spectra in Figure 2.96 are isomeric compounds. Identify them. You should be able to differentiate the isomers.

PROBLEM 2.36

The mass spectrum in Figure 2.97 is of a compound that has a typical indole spectrum in the ultraviolet. What is the compound?

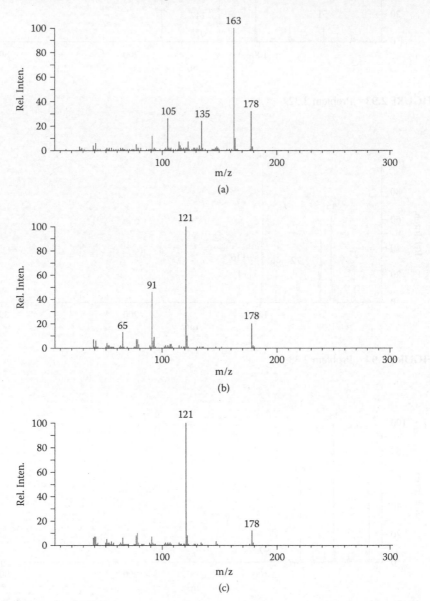

FIGURE 2.96 Problem 2.35.

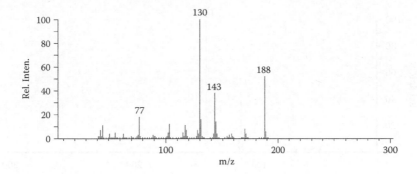

FIGURE 2.97   Problem 2.36.

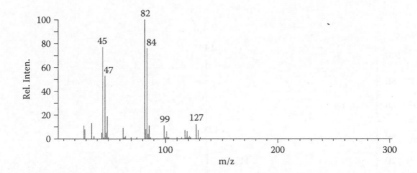

FIGURE 2.98   Problem 2.37.

PROBLEM 2.37
Identify the compound that gives the mass spectrum in Figure 2.98. The infrared spectrum contains an acid carbonyl band.

PROBLEM 2.38
The mass spectra in Figures 2.99 and 2.100 are of 3-methylquinoline and 7-methylquinoline, respectively. Identify the compounds giving the spectra in Figures 2.101 and 2.102.

3-Methylquinoline   7-Methylquinoline

PROBLEMS 2.39–2.51
Identify the compounds that give the spectra in Figures 2.103 to 2.115 (Problems 2.39–251, respectively).

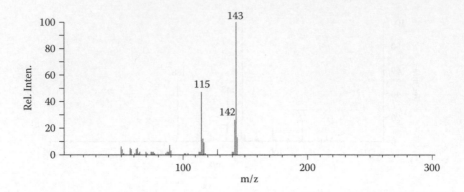

FIGURE 2.99 Problem 2.38.

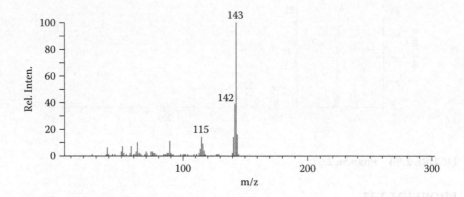

FIGURE 2.100 Problem 2.38.

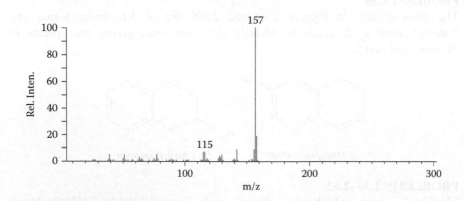

FIGURE 2.101 Problem 2.38.

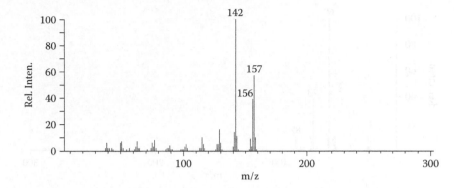

FIGURE 2.102   Problem 2.38.

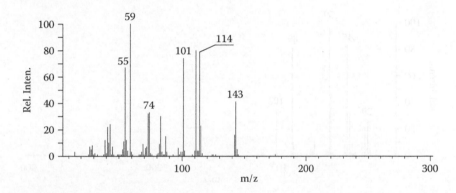

FIGURE 2.103   Problem 2.39.

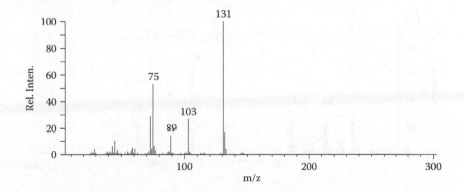

FIGURE 2.104   Problem 2.40.

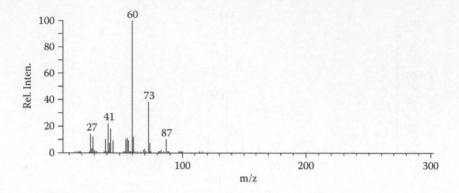

FIGURE 2.105   Problem 2.41.

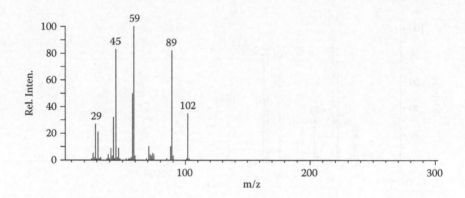

FIGURE 2.106   Problem 2.42.

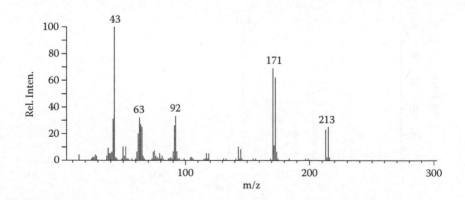

FIGURE 2.107   Problem 2.43.

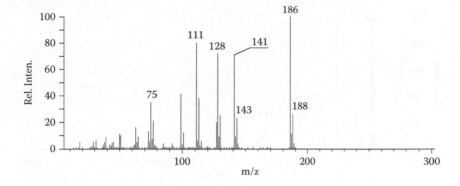

FIGURE 2.108 Problem 2.44.

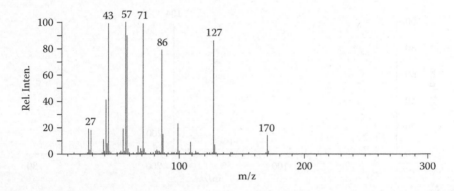

FIGURE 2.109 Problem 2.45.

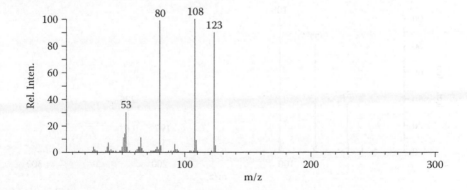

FIGURE 2.110 Problem 2.46.

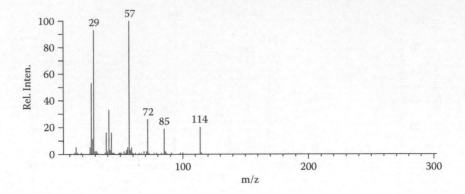

FIGURE 2.111   Problem 2.47.

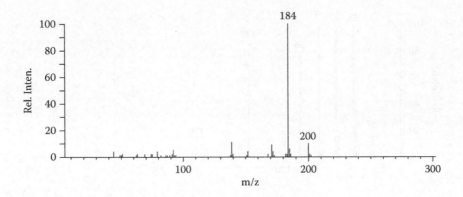

FIGURE 2.112   Problem 2.48.

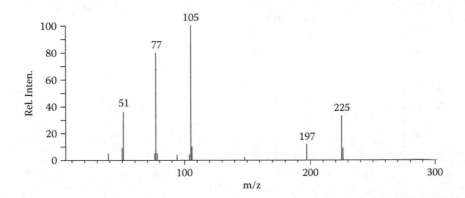

FIGURE 2.113   Problem 2.49.

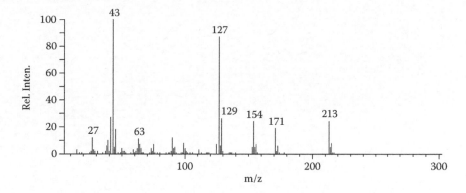

FIGURE 2.114   Problem 2.50.

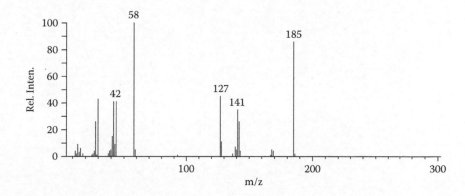

FIGURE 2.115   Problem 2.51.

3 Accurate Mass Measurement

INTRODUCTION

Chapters 1 and 2 introduced the various processes (ionization and fragmentation) that occur in the mass spectrometer and the means by which they can be used for reconstructing a molecular structure from the observed fragments. The nominal molecular weight and masses of fragment ions, when supplemented by other spectroscopic information, can often lead to the solution of an organic structure problem. One must, however, rely on making the correct assumptions concerning the compositions of the various ions. In many cases, these assumptions are quite straightforward. For example, a hydrocarbon that lacks strong absorption in the ultraviolet (UV) band and has a molecular weight of 182 must have an elemental composition of $C_{13}H_{26}$ (one ring or double bond). The alternative $C_{14}H_{14}$ composition would be easily recognized in the mass spectrometer as either a polyene or polycyclic compound and would probably be UV absorbing.

PROBLEM 3.1
Determine the elemental composition of each of the following compounds.

1. A substance with both aromatic and carbonyl absorption in the infrared and a molecular weight of 136.
2. A compound with a strong molecular ion peak at m/z 144 and only weak fragment peaks. This material becomes water soluble under acidic conditions.
3. This compound exhibits a three-proton singlet in its nuclear magnetic resonance (NMR) spectrum at $\delta = 3.6$, typical for a methyl ester. Its molecular weight is 104, and a strong fragment peak is seen at m/z 86.
4. This compound has a molecular weight of 206 (with a $P + 2$ isotope peak of equal intensity) and has no aliphatic proton signals in the NMR (there is a complex signal between $\delta = 6.7$ and $\delta = 7.7$).

The examples in Problem 3.1 are provided to remind you of the importance of additional structural information that might be available. Even with no knowledge of fragmentation reactions and interpretation techniques, a considerable amount of information can be extracted from a spectrum. Determination of the compositions, and even structures, of these compounds should offer no difficulty.

Elemental compositions may also be deduced in certain cases by comparing a spectrum of an unknown substance with those of model compounds. As the examples at the end of Chapter 2 demonstrated, a small difference in molecular weights of related compounds is often attributable to only one additional atom or functional group.

- Homologs usually give identical (or nearly so) mass spectra, except that the peaks are displaced by 14 mass units.
- A methoxy group on one of two related compounds will manifest itself in a mass shift of 30 units (CH_3O versus H).

We have already seen, however, situations in which elemental compositions are difficult to determine by conventional mass spectrometry. This is particularly true in the field of natural products chemistry, in which only submilligram amounts of material are isolated, and mass spectrometry is often the sole source of structural information. It is here that the development of high-resolution mass spectrometry has opened exciting new areas of exploration. High-resolution mass spectrometry, through the accurate measurement of ion masses, gives direct evidence concerning the elemental compositions of ions.

Aston discovered in 1923 that every isotopic mass is characterized by a small "mass defect"; that is, the masses of nuclides are not simple multiples of a fundamental unit. Since Aston's discovery, considerable work has been performed to measure accurately the masses of all the known stable nuclides (mainly by the use of mass spectrometers). Table 1.1 (in Chapter 1) contains the results of this work for the elements commonly found in organic compounds.

The accepted atomic weights for the elements are included to point out the difference between the conceptions of the chemist and of the mass spectroscopist regarding the mass of an atom. Since the chemist is accustomed to using atomic weights, which are average weights of all the isotopes of an element, the chemist is sometimes confused by an accurate mass as determined by mass spectrometry. Remember that heavy isotopes do not contribute to the mass of an ion but result in altogether different ions at nominally different masses.

For example, the molar mass of acetone (C_3H_6O) is 58.0707, but the molecular ion, since it is monoisotopic in each element, has an exact mass of $(3*12) + (6*1.0078) + 15.9949 = 58.0417$.

PROBLEM 3.2

Calculate the exact mass of the ^{13}C isotope ion of acetone.
Calculate the exact mass of the ^{17}O isotope ion.

Butane is a nominal isobar of acetone, but its molecular ion has an exact mass of $(4*12) + (10*1.0078) = 58.0783$. The difference in mass between butane and acetone is 0.0364 Da. They can be completely separated (less than 10 percent peak overlap) from one another by a mass spectrometer with a resolution of 1,600, as shown by Equation 3.1:

$$R = \frac{58}{0.0364} = 1600 \tag{3.1}$$

Furthermore, a mass measuring accuracy of only ±0.02 Da would be required for unambiguous differentiation of the two ions.

TABLE 3.1

Possible Combinations of Carbon, Hydrogen, Nitrogen, and Oxygen for an Ion of Mass 146

Composition	Exact Mass	Composition	Exact Mass
$C_4H_2O_6$	145.9851	C_9H_8NO	146.0606
$C_8H_2O_3$	146.0004	$C_5H_{10}N_2O_3$	.0691
$C_4H_4NO_5$	.0089	$C_8H_8N_3$	.0718
$C_{12}H_2$	.0157	$C_{10}H_{10}O$	.0732
$C_7H_2N_2O_2$	.0116	$C_4H_{10}N_4O_2$	.0804
$C_5H_6O_5$	.0215	$C_9H_{12}NO_3$	.0817
$C_6H_2N_4O$	.0229	$C_9H_{10}N_2$	.0844
$C_8H_4NO_2$	.0242	$C_5H_{12}N_3O_2$	.0929
$C_4H_6N_2O_4$	.0328	$C_7H_{14}O_3$	.0943
$C_5H_2N_6$	.0341	$C_{10}H_{12}N$	.0970
$C_7H_4N_3O$	.0354	$C_4H_{12}N_5O$	.1042
$C_9H_6O_2$	.0368	$C_6H_{14}N_2O_2$	.1055
$C_5H_8NO_4$	.0453	$C_{11}H_{14}$	.1095
$C_4H_4N_5$	.0467	$C_5H_{14}N_4O$	.1168
$C_8H_6N_2O$	.0480	$C_7H_{16}NO_2$	.1181
$C_4H_8N_3O_3$	.0566	$C_4H_{14}N_6$	.1280
$C_6H_{10}O_4$	.0579	$C_8H_{18}O_2$	.1307
$C_7H_6N_4$	.0592		

Beynon was among the first to recognize that exact ion masses can be used to determine elemental compositions of the ions.

To illustrate, the molecular ion of a compound was measured with an accuracy of 0.001 Da (a typical accuracy) to be 146.0738. Table 3.1 contains all possible combinations of carbon, hydrogen, nitrogen, and oxygen (with a maximum of six nitrogen or oxygen atoms) for an ion of nominal mass 146.

Which is the correct composition?
Is there more than one possible composition, in view of the accuracy of the measurement?

Note that resolution of the ions in this list would be extremely difficult: The ions $C_4H_8N_3O_3$ and $C_6H_{10}O_4$ require a resolving power of 100,000 for complete separation. Fortunately, since these compositions are so widely different, such a doublet is unlikely in the spectrum of a pure compound. The mass-measuring accuracy of an instrument is considerably more important than its resolution, although both parameters are usually closely related.

If the other seven elements of Table 1.1 are considered in the example, the list of possible compositions grows substantially; there are over 600 possible combinations of the 11 elements that result in a nominal mass of 146. Many times, however, accurate mass measurements can be supplemented by isotope peak observations. The composition $C_7H_{13}NCl$, for example, has an exact mass of 146.0737, but the $P + 2$ isotope peak characteristic of a chlorine atom would easily distinguish it from $C_{10}H_{10}O$, the correct answer to the preceding example.

The advantages of high-resolution mass spectrometry or, more correctly, the accuracy of the masses derived from it, are obvious. But before consideration of their applications, a discussion of the techniques for mass calibration and elemental composition calculation is in order.

MASS CALIBRATION

Accurate mass scales (or calibration curves) are generally established by measuring the mass spectrum of a reference compound simultaneously with the spectrum of the sample. The precise mass of every ion in the spectrum of the reference compound is known, so a precise mass correlation is thereby provided. Common reference materials are perfluorokerosene (PFK) and perfluorotributylamine (PFTBA), the mass spectra of which are shown in Figures 3.1 and 3.2, respectively. Since all the ions formed from these compounds contain several fluorine atoms (18.9984) and no hydrogen atoms (1.0078), they have negative mass defects and are well separated from organic ions that normally have positive mass defects. Of course, other chemicals may be used to provide reference masses, as long as the exact masses in its spectrum are known.

Figure 3.3 shows a small portion of a mass spectrum of PFK and an organic compound. The most intense peak near mass 169 in Figure 3.3a is from the PFK. The other smaller peaks are from an unknown chemical. The peaks are centroided, and a new calibration table is generated from PFK ions in the spectrum. The accurate masses can then be determined, as shown in Figure 3.3b. The peak labeled with 169.0897 is shown to be $C_{12}H_{11}N$, which has an exact calculated mass of 169.0891. The smaller peak has a composition of $C_{11}H_7ON$, which has an exact calculated mass of 169.0528.

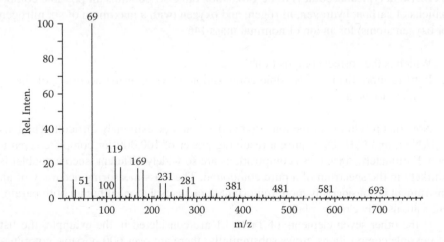

FIGURE 3.1   Mass spectrum of perfluorokerosene (PFK).

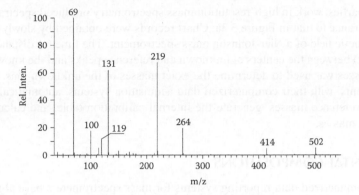

FIGURE 3.2 Mass spectrum of perfluorotributylamine (PFTBA).

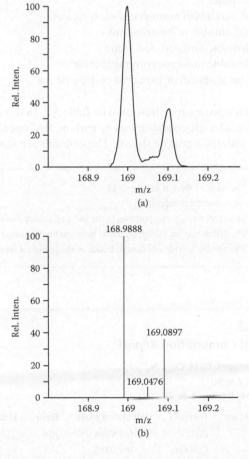

FIGURE 3.3 High-resolution (a) raw data spectrum and (b) accurate masses calculated from internal calibration table.

The earliest work in high-resolution mass spectrometry produced spectra similar in appearance to that in Figure 3.3a. Chart records were obtained by slowly varying the magnetic field of a Nier-Johnson mass spectrometer. The time (or distance along the chart) between the centers of unknown and reference peaks and the known reference masses was used to determine the exact masses of the unknown ions. Current instruments, with their computerized data acquisition systems, automatically identify the reference masses, generate the internal calibration table, and calculate the accurate masses.

ELEMENTAL COMPOSITIONS

Most computerized data-reporting systems for mass spectrometers have algorithms that will rapidly calculate all possible elemental compositions for a given ion mass. Of course, some limitations are normally set with regard to

- Elements to consider
- Minimum and maximum numbers of each element
- Maximum total number of heteroatoms
- Maximum allowable unsaturation value
- Maximum allowable mass measurement error
- The presence (or absence) of expected isotope peaks

The output of such a program is illustrated in Table 3.2. In this case, the computer was instructed to consider all combinations of carbon, hydrogen, nitrogen, oxygen, and fluorine atoms with the limits as shown. The columns are read as follows:

Measured Mass	The mass of the ion as measured
Formula	An elemental composition
Calculated Mass	The exact mass corresponding to the indicated composition
Error	The difference (in millimass units) between the measured and calculated mass
Unsaturation	The number of rings or double bonds in the indicated formula

TABLE 3.2

Elemental Composition Report

Selected isotopes: C, H, O_{0-3}, N_{0-3}, F_{0-10}
Error limit: 2 mmu
Unsaturation limits: −1 to 20

Measured Mass	Formula	Calculated Mass	Error	Unsaturation
168.9888	C_3F_7	168.9888	0	0.5
169.0476	$C_9H_7OF_2$	169.0465	1.1	5.5
169.0897	$C_{12}H_{11}N$	169.0891	0.6	8

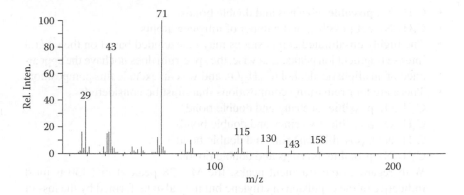

FIGURE 3.4 Exercise 3.1.

APPLICATIONS

High-resolution mass measurements and the subsequent elemental compositions of molecular and fragment ions can indeed be valuable aids in the interpretation of mass spectra. In many instances, the unambiguous empirical formula of a compound and a low-resolution spectrum showing the characteristic fragmentation pattern can be interpreted in terms of a single structure. In more difficult problems, elemental compositions of fragment ions offer evidence concerning which atoms are involved in the fragmentation reactions and thereby lead to a structure. For example, the low-resolution spectrum in Figure 3.4 can be interpreted in several ways since the relationships between the various peaks are ambiguous. The high-resolution data, however, provide conclusive evidence for the empirical formula and ultimately the structure of this compound. It is not difficult to rationalize each peak once the structure is known.

EXERCISE 3.1

Interpret the spectrum in Figure 3.4 in terms of a structure.

The peak at m/z 158 fails no test for a molecular ion. If this is indeed the molecular ion, there are several possible elemental compositions to be considered (we neglect atoms other than C, H, N, and O).

- $C_{11}H_{26}$: not possible, too many hydrogen atoms
- $C_{12}H_{14}$: **possible**, six rings and double bonds
- $C_{11}H_{10}O$: **possible**, seven rings and double bonds
- $C_{10}H_{22}O$: **possible**, no rings and double bonds
- $C_{10}H_{24}N$: not possible, odd number of nitrogen atoms
- $C_9H_{18}O_2$: **possible**, one ring and double bond
- $C_9H_{20}NO$: not possible, odd number of nitrogen atoms
- $C_8H_{14}O_3$: **possible**, two rings and double bonds
- $C_8H_{18}N_2O$: **possible**, one ring and double bond
- $C_8H_{16}NO_2$: not possible, odd number of nitrogen atoms

- $C_9H_{22}N_2$: **possible**, no rings and double bonds
- $C_8H_{20}N_3$: not possible, odd number of nitrogen atoms
- The highly unsaturated compositions may be excluded based on the several intense fragment ion peaks. Likewise, the spectrum does not have the appearance of an aliphatic alcohol ($C_{10}H_{22}O$), and we can exclude this composition.
- There are four remaining compositions that must be considered:
  $C_9H_{18}O_2$: **possible**, one ring and double bond
  $C_8H_{14}O_3$: **possible**, two rings and double bonds
  $C_8H_{18}N_2O$: **possible**, one ring and double bond
  $C_9H_{22}N_2$: **possible**, no rings and double bonds
- With regard to the fragment peaks, the M - 28 peak at m/z 130 is most indicative of the expulsion of ethylene but may also be formed by the loss of a molecule of carbon monoxide. The remaining fragment ions are equally ambiguous, but the high-resolution data in Table 3.3 clarify every question and transform this into a relatively simple problem.
- The empirical formula of the compound is $C_8H_{14}O_3$ with two rings or double bonds. The M - C_2H_4 ion (m/z 130) along with the M - OC_2H_5, ion (m/z 113) is characteristic of an ethyl ester function. This is corroborated by the ion at m/z 88 ($C_4H_8O_2$), which corresponds to the McLafferty rearrangement product.

m/z 88

TABLE 3.3
Elemental Compositions of Ions in the Spectrum of Figure 3.4

Exact Mass	Composition	Relative Intensity	Exact Mass	Composition	Relative Intensity
41.0391	C_3H_5	17.7	61.0289	$C_2H_5O_2$	4.0
42.0106	C_2H_2O	15.5	69.0340	C_4H_5O	20.0
42.0470	C_3H_6	4.5	70.0418	C_4H_6O	6.2
43.0184	C_2H_3O	55.4	71.0497	C_4H_7O	100.0
43.0548	C_3H_7	16.6	84.0211	$C_4H_4O_2$	11.8
43.9898	CO_2	5.4	87.0446	$C_4H_7O_2$	18.0
44.0626	C_3H_8	0.6	88.0524	$C_4H_8O_2$	11.7
44.9977	CHO_2	8.8	113.0603	$C_6H_9O_2$	20.5
45.0340	C_2H_5O	5.2	115.0395	$C_5H_7O_3$	44.2
51.0235	C_4H_3	2.8	130.0630	$C_6H_{10}O_3$	13.3
54.0106	C_3H_2O	6.0	143.0708	$C_7H_{11}O_3$	3.9
60.0211	$C_2H_4O_2$	9.8	158.0943	$C_8H_{14}O_3$	10.4

- The fragment at m/z 115($C_5H_7O_3$) shows that a propyl chain is present in a position where it is easily lost by simple cleavage. The butyryl ion ($C_4H_7O_2$) at m/z 71, the base peak in the spectrum, demonstrates the mode of attachment of the three-carbon chain.

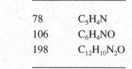

m/z 71

- It is a simple matter to now add the masses of the known portions of the molecule to find that the entire structure has been described. There is no remaining mass that has to be accounted for.

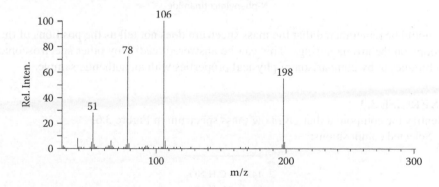

71 + 87 = 158

The following exercises include both high-resolution and low-resolution data. In each case, it would be useful practice to attempt a solution without referring to the high-resolution data, turning to them only when difficulties are encountered.

EXERCISE 3.2
Identify the compound that gives the mass spectrum in Figure 3.5.
   Selected compositions are

78	C_5H_4N
106	C_6H_4NO
198	$C_{12}H_{10}N_2O$

FIGURE 3.5   Exercise 3.2.

- The ion $C_{12}H_{10}N_2O$ fails no test for a molecular ion. This composition indicates there are nine rings or double bonds in the molecule, a fact supported by the small number of fragment peaks. The fragment ion $C_5H_4N^+$ is characteristic of a pyridine moiety, and attached to it is a carbonyl function, as evidenced by the ion $C_6H_4NO^+$.

m/z 106

- By subtraction, the remainder of the molecule contains the elements C_6H_6N and could have only two possible structures: one constituting an N-phenyl amide function (A) and the other an aminophenyl functional group (B).

A   B

- The latter can be ruled out since an aminobenzoyl ion (m/z 120) would be expected from such a compound.
- The correct structure is therefore N-phenylnicotinamide

N-phenylnicotinamide

It should be remembered that the mass spectrum does not tell us the positions of the groups on the aromatic rings. This can be answered readily by other spectroscopic techniques or by comparison of physical properties with an authentic sample.

EXERCISE 3.3

Identify the compound that gives the mass spectrum in Figure 3.6.

Selected compositions:

143	$C_6H_9NO_3$
84	C_4H_6NO

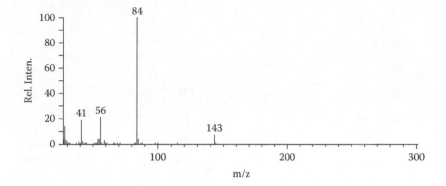

FIGURE 3.6   Exercise 3.3.

- The ion $C_6H_9NO_3$ is a good candidate for the molecular ion. There are three rings or double bonds in this molecule.
- The even-electron ion $C_4H_6NO^+$ supplies the clue to the structure of this compound. It is formed by the loss of a $C_2H_3O_2$ fragment, which from Table 2.7 is characteristic of a methyl ester. Of the possible structures for the remainder of the molecule (that forming the intense ion at m/z 84), the cyclic lactam is a likely candidate. Open-chain, unsaturated structures would be expected to fragment more extensively, and the loss of the $C_2H_3O_2$ neutral fragment would not be so prevalent if it were attached to the ring at another position.

$$m/z\ 143 \longrightarrow m/z\ 84\ +\ \cdot COOCH_3$$

EXERCISE 3.4
Identify the compound that gives the spectrum in Figure 3.7.
   Selected compositions:

180	$C_{11}H_{16}S$
123	C_7H_7S
110	C_6H_6S

- You should have been able to solve this problem without referring to the accurate mass data. The presence of sulfur is indicated by the $P+2$ isotope peak of the ion at m/z 180, which appears to be the molecular ion.

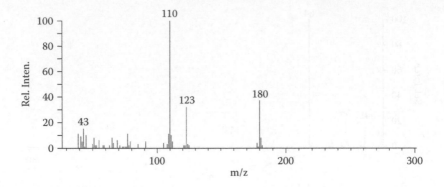

FIGURE 3.7   Exercise 3.4.

- The very strong odd-electron ion at *m/z* 110 is formed by the elimination of 70 mass units (pentene, C_5H_{11}) in a McLafferty-type rearrangement. Thus, there must be a five-carbon chain one atom removed from the benzene ring, and by subtraction that atom must be sulfur. We cannot state with certainty the nature of the carbon chain, but the absence of cleavages to give higher-mass fragment peaks supports an unbranched structure.

PROBLEMS 3.3 TO 3.6

Identify the compounds that give the mass spectra in Figures 3.8 to 3.11 (for Problems 3.3–3.6, respectively). Elemental compositions for selected ions are given in the following:

For Figure 3.8:

Nominal Mass	Formula from Accurate Mass
124	$C_8H_{12}O$
81	C_5H_5O

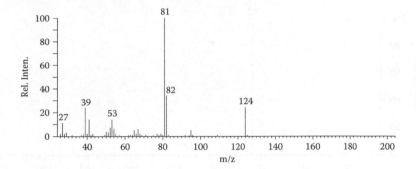

FIGURE 3.8 Problem 3.3.

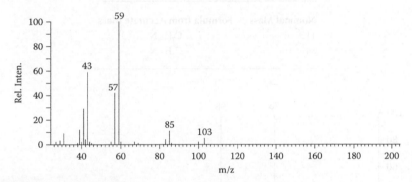

FIGURE 3.9 Problem 3.4.

For Figure 3.9:

Nominal Mass	Formula from Accurate Mass
103	$C_5H_{11}O_2$
85	C_5H_9O
59	C_3H_7O

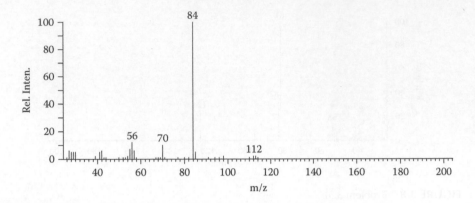

FIGURE 3.10   Problem 3.5.

For Figure 3.10:

Nominal Mass	Formula from Accurate Mass
112	$C_7H_{14}N$
84	$C_5H_{10}N$

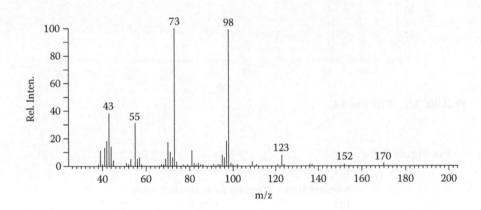

FIGURE 3.11   Problem 3.6.

For Figure 3.11:

Nominal Mass	Formula from Accurate Mass
170	$C_9H_{14}O_3$
123	$C_7H_7O_2$
98	$C_5H_6O_2$
73	C_4H_9O

4 Modern Techniques

INTRODUCTION

You should now be at least somewhat versed in the interpretation of mass spectra in terms of fragmentation reactions and how the products of these reactions (the ions) are used to reconstruct molecular structures. The chemistry of mass spectrometry has been presented to relate many of the fragmentation processes to one another and to the more familiar solution chemistry. When taken in this light, mass spectra can usually be interpreted by the individual chemist, so that the chemist does not have to rely on an experienced mass spectroscopist. We have assumed in our discussions, however, that (1) the spectrum of a compound can be obtained with little or no difficulty by conventional means, and (2) the compound fragments in a manner characteristic of its structure. In the more routine cases, with fairly large (>1 mg) amounts of a relatively volatile sample, spectra are obtained by introducing the material through a reservoir. Somewhat less volatile and smaller samples may be introduced through a direct insertion probe into the ion source itself. Many problems are encountered, however, that require special treatment, particularly when a material is very polar and does not have sufficient vapor pressure.

This chapter describes a few of the means by which some of the more difficult problems can be solved by mass spectrometry. Many simple chemical derivatives of polar compounds, particularly those containing –NH or –OH groups that form hydrogen bonds, can be prepared to increase volatility. At the same time, these derivatives introduce changes in molecular weights and fragmentation patterns that can be useful for interpreting the spectra. Isotope labeling has long been used as a tool for elucidation of structures and fragmentation pathways.

Special operating techniques may yield the additional information required for solving a problem. One may desire to minimize fragmentation by using a less-energetic method for producing molecular ions. Thermal effects in a spectrum may be reduced by operating the ion source or inlet system at lower temperatures. Last, special introduction techniques may be required to obtain a usable mass spectrum. Two important advances in sample introduction techniques utilize a gas chromatograph or a liquid chromatograph with the effluent passing directly into the ion source of the mass spectrometer.

GAS CHROMATOGRAPHY–MASS SPECTROMETRY

INTRODUCTION

Gas chromatography has long been recognized as an efficient method for separating and even identifying volatile organic compounds. With the advent of highly specific capillary columns and temperature-programmed ovens, gas chromatography has been extended into many areas of organic chemistry, including the natural products field. Such polar compounds as steroids, amino acids, alkaloids, and sugars have been adapted for gas chromatographic separation. Early in the development of organic mass spectrometry, it was felt that gas chromatography could play an important role in purifying samples for subsequent mass spectral analysis. Both techniques require only minute amounts of sample and depend on at least a small degree of volatility.

The separation of materials afforded by a gas chromatograph can be used to purify compounds that are difficult to purify by other means. Each component of a mixture passes into the mass spectrometer essentially as a pure substance.

The advantages of the gas chromatograph-mass spectrometer (GCMS) over either of the instruments in isolated use are quite distinct. Not only are the various components of a mixture separated, but direct structural evidence of each component is provided through the mass spectra. In addition, mass spectra can be determined with very small amounts of sample.

Figure 4.1 shows a typical chromatogram from a GCMS analysis. The chromatogram, termed the total ion chromatogram (TIC), is reconstructed from a sum of all ion intensities for each individual mass spectrum. The x-axis represents time, in minutes, from the sample injection.

There may be several hundred, or even thousand, mass spectra associated with a single GCMS analysis. Thus, a computerized system is required for collecting and reviewing the data. Such a system offers unlimited avenues for quickly evaluating the data for both qualitative (i.e., identification) and quantitative (compound concentration) purposes.

For example, the data shown in Figure 4.1 can be scanned quickly to detect all compounds in the alkyl benzene family, which always give a strong fragment at 91 Da (Figure 4.2).

In this case, four compounds are quickly identified by this technique. The mass spectrum corresponding to the peak eluting at 8.99 minutes is shown in Figure 4.3.

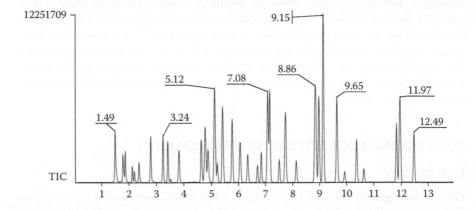

FIGURE 4.1   Total ion chromatogram (TIC) from GCMS analysis.

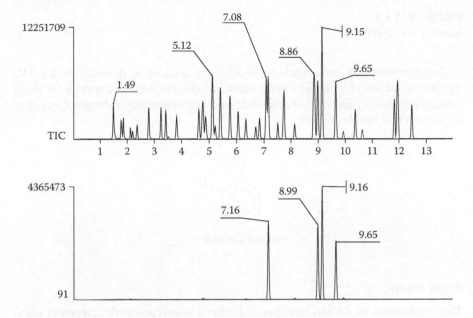

FIGURE 4.2 Reconstructed ion chromatogram (RIC) for mass 91.

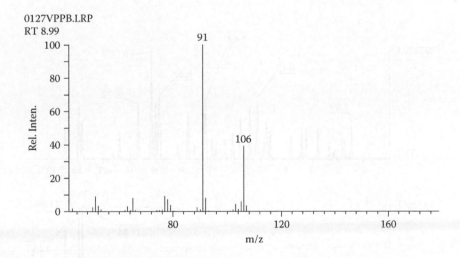

FIGURE 4.3 Mass spectrum at 8.99 minutes.

PROBLEM 4.1
Identify this compound.

Another series of compounds related to one another is detected by an RIC (reconstructed ion chromatogram) for mass 146 (Figure 4.4). The spectra for these components are all identical (Figure 4.5) and are indicative of dichlorobenzenes, with molecular weights of 146.

Dichlorobenzene

APPLICATIONS

The applications of GCMS have generally been concerned with compound identification in complex mixtures. The high sensitivity of the mass spectrometer can also lead to detection and identification of compounds present at concentrations below the detection limit of other analytical techniques. The TIC chromatogram in Figure 4.6 was obtained from analysis of the headspace above a pharmaceutical product.

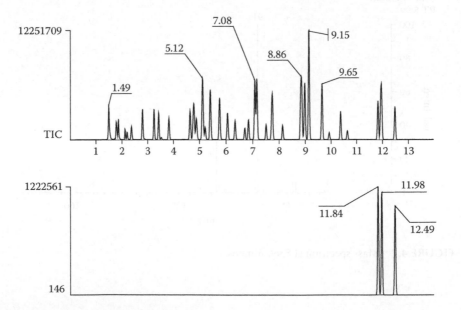

FIGURE 4.4 TIC and RIC for mass 146.

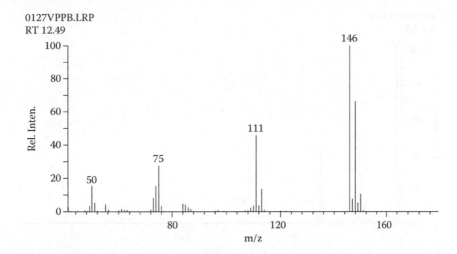

FIGURE 4.5   Mass spectrum at 12.49 minutes.

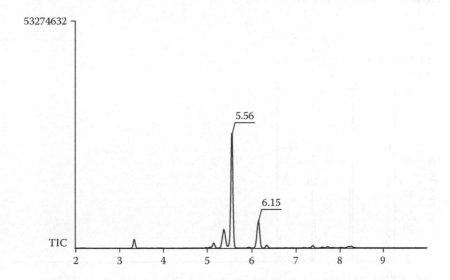

FIGURE 4.6   TIC from GCMS analysis of headspace gases above a pharmaceutical product.

EXERCISE 4.1

Identify the compounds corresponding to the two strongest peaks in Figure 4.6 at retention times (RT) 5.5 minutes and 6.15 minutes. The mass spectra are shown in Figures 4.7 and 4.8.

The mass spectrum in Figure 4.7 contains what appears to be a molecular ion at m/z 100 and typical alkane fragments at m/z 43, 57, and 71. With no particular fragment indicating preferred fragmentation, this spectrum corresponds to n-heptane (C_7H_{16}, $M = 100$).

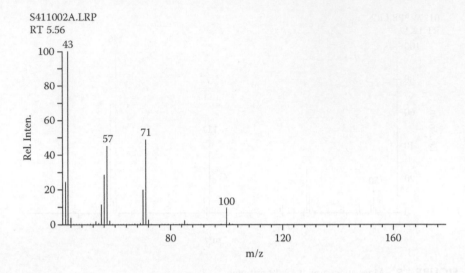

FIGURE 4.7   Mass spectrum at RT 5.56.

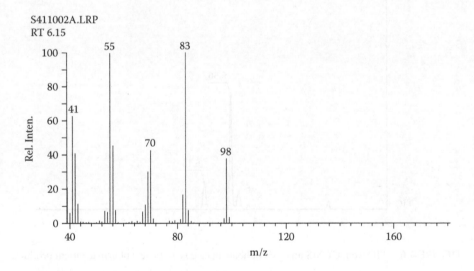

FIGURE 4.8   Mass spectrum at RT 6.15.

The mass spectrum in Figure 4.8 has a very different appearance. An apparent molecular ion is at m/z 98. This corresponds to a C_7H_{14} compound, with one ring or double bond ($M = 98$). The relatively strong molecular ion is indicative of a cyclic structure, and the very strong M−15 (M−CH_3) fragment indicates the presence of a methyl function. This spectrum corresponds to methylcyclohexane.

Methylcyclohexane

Another application of GCMS is in obtaining partial separation by the gas chromatograph of such similar materials as isotopically labeled compounds and analysis of the gas chromatographic peak by the mass spectrometer. The TIC chromatogram in Figure 4.9 shows only a single peak. However, the RICs for dichlorobenzene ($M = 146$) and D4-dichlorobenzene ($M = 150$) reveal a slight chromatographic resolution between these closely related compounds. This observation is particularly important for making quantitative measurements.

LOW-ENERGY IONIZATION

In the majority of mass spectra examined in previous chapters, a molecular ion peak was present with an intensity that made it relatively easy to identify. In some cases, however, the molecular ion was either absent or of extremely low intensity; the molecular ion was too unstable to withstand the high energy imparted to it by the electron beam. The remainder of this chapter is devoted to descriptions of the various methods that have been devised to produce molecular ions in lower-energy states.

- Low-electron energy
- Field ionization (FI)

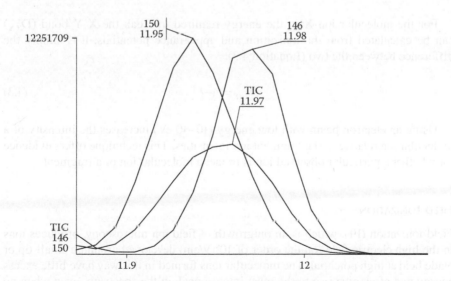

FIGURE 4.9 Overlapped chromatograms for TIC and RICs for m/z 146 and m/z 150.

- Fast atom bombardment (FAB)
- Photoionization (PI)
- Chemical ionization (CI)
- Electrospray ionization (ESI)
- Atmospheric pressure chemical ionization (APCI)
- Matrix-assisted laser desorption ionization (MALDI)
- Direct analysis in real time (DART)

LOW ELECTRON ENERGY

It was stated in Chapter 2 that organic molecules in the vapor state are ionized when an amount of energy equal to the ionization potential is transferred from an electron beam to the molecules, as depicted in Equation 4.1:

$$XY \xrightarrow{\text{electron energy } I} XY^+ + e^- \tag{4.1}$$

The *ionization potential I* can be measured by mass spectrometry. The minimum electron energy at which the molecular ion is formed is determined by lowering the potential between the filament and electron trap in small steps until the molecular ion is no longer detected.

If the energy transferred to the molecules is much greater than the ionization potential, fragmentation occurs. There is a minimum energy that must be supplied to form each fragment ion from a molecule. This is termed the *appearance potential A* of the ion and can be measured in the same manner, as depicted in Equation 4.2:

$$XY \xrightarrow{\text{electron energy } A} X^+ + Y + e^- \tag{4.2}$$

For the molecular ion XY^+, the energy required to break the X–Y bond (D_{X-Y}) can be calculated from the ionization and appearance potentials; it is simply the difference between the two (Equation 4.3):

$$D_{X-Y} = A - I \tag{4.3}$$

Using an electron beam with low energy (10–30 eV) increases the intensity of a molecular ion relative to the fragment ion intensities. This technique offers evidence for whether a particular observed ion is in fact a molecular ion or a fragment.

FIELD IONIZATION

Field ionization (FI), which is an outgrowth of field ion microscopy, produces ions in the high electric field (on the order of 10^8 V/cm) developed around a small tip or blade held at high potential. The molecular ions formed in this way have little excess energy and often give rise to the most intense peak in the spectrum, even when no molecular ion is seen in the spectrum produced by electron bombardment.

This phenomenon is of particular importance in the analysis of complex mixtures of organic compounds (such as those obtained from petroleum distillates); the absence of fragment ion peaks makes identification and measurement of molecular ion abundances easier and, perhaps of greater significance, more reliable. The advantages of an unambiguous identification of the molecular ion in structure elucidations have already been discussed.

In FI, a high-potential electric field is applied to an emitter with a sharp surface, such as a razor blade or a filament from which tiny "whiskers" have formed. This results in a very high electric field, which can result in ionization of gaseous molecules of the analyte. Mass spectra produced by FI have little or no fragmentation. They are dominated by molecular radical cations M^+ or, less often, by protonated molecules $[M + H]^+$.

Fast Atom Bombardment

FAB is a relatively soft ionization technique using a high-energy (4,000- to 10,000-eV) beam of atoms, typically argon or xenon. The material to be analyzed is mixed with a nonvolatile matrix chemical. Common matrices include glycerol, thioglycerol, 3-nitrobenzyl alcohol (3-NBA), 18-crown-6 ether, 2-nitrophenyloctyl ether, sulfolane, diethanolamine, and triethanolamine.

FAB produces primarily intact protonated molecules denoted as $[M + H]^+$ in positive ion spectra and deprotonated molecules such as $[M - H]^-$ in negative ion spectra. The nature of its ionization products places it close to ESI and MALDI.

Photoionization

A beam of high-energy photons (wavelengths less than 1,200 Å have energies greater than 10 eV) may also be used to promote ionization of organic molecules in the vapor state. As with the field ion source (and with low-voltage electron ionization), the resulting molecular ions contain little excess energy, and fragmentation is minimal. The PI source has another important feature: With the use of a vacuum monochromator, an energetically well-defined (± 1 Å $= \pm 0.008$ eV) ionizing beam can be obtained over a fairly broad range of energies. Since electron beams generally have energy spread in excess of ± 0.5 eV, PI is particularly useful for determining accurate ionization potentials. Close examination of the PI efficiency curves may yield valuable information concerning the electronic and vibrational energy levels involved in ionization and fragmentation.

The molecular weight and analytical applications mentioned for the field ion source can be extended to include this source. Further, proper choice of wavelength may be made so that only single components in a mixture are ionized. For this technique to be successful, of course, the ionization potentials must be separated by 2 to 3 eV; otherwise, the various components might interfere with each other.

Many different photon sources have been tried, among them electrical discharges in hydrogen or the inert gases and the newer laser beams. Although typical photon beams have rather low energies, an emission line of helium at 304 Å is nearly as energetic (40.8 eV) as the electron beams. Mass spectra produced by this high-energy photon beam are remarkably like electron bombardment spectra.

Physically, the PI source resembles the electron bombardment source, with a light source and monochromator replacing the heated filament and electron trap. Since there is no heated filament, the PI source has the further advantage that decompositions promoted by heat are eliminated.

PI mass spectra usually are dominated by molecular radical cations M^+ or, less often, by protonated molecules $[M + H]^+$.

CHEMICAL IONIZATION

High-pressure mass spectrometry has been studied for many years by workers interested in ion-molecule reactions. Ionization of the substance under scrutiny is affected by reactions between the molecules of the compound and a set of ions that serve as ionizing reactants. The reactant ions are formed by subjecting a gas to electron bombardment at pressures of about 1 mm. The reactant gas is extensively ionized and, because of the high pressure, undergoes ion molecule reactions with itself and with the sample molecules that are present at relatively low concentrations (less than 1 percent. This process in a mass spectrometer is termed chemical ionization CCI).

If the reactant gas is methane, the most abundant ions formed are CH_5^+ and $C_2H_5^+$, which then become the reactant ions. On collision with the sample molecules, these ions donate a proton to form what is termed a *quasi-molecular ion* **[M + H]⁺** at a mass 1 Da higher than the molecular weight (Equation 4.4).

$$M + CH_5^+ \rightarrow MH^+ + CH_4 \qquad (4.4)$$

There are two important distinctions between the quasi-molecular ion and a molecular ion formed by conventional electron. First, the energy imparted to the sample molecule during the collision-induced ionization is relatively low (<20 eV for methane as the reactant gas). Second, the quasi-molecular ion is an even-electron species and therefore more stable than the odd-electron molecular ion. Quasi-molecular ions are therefore almost always seen in CI mass spectra.

Unlike the previous low-energy forms of ionization, however, CI produces fragment ions that can be used for determining structures. Elimination fragmentations are among the most common under these conditions, and other rearrangement fragmentations are seldom observed. The spectra in Figure 4.8 reflect the differences between electron bombardment and CI mass spectra of a complex organic compound and the way in which the information extracted from each complements the other.

O-methylpellotine

The EI spectrum of O-methylpellotine ($M = 251$) in Figure 4.10a reflects a problem that is encountered in the interpretation of the mass spectra of tetrahydroisoquinoline alkaloids in general. A molecular ion peak is not observed, but an intense peak in the high-mass region occurs at m/z 236, and a weak peak is seen 14 mass units higher at m/z 250. There are two possible interpretations for these data.

1. The ion of mass 236 may correspond to an M − H ion, and a small homolog impurity would account for the m/z 250 peak.
2. Alternatively, the higher-mass ion might be an M − H ion, and the m/z 236 peak would then correspond to M − CH_3.

The quasi-molecular ion [M + H]$^+$ at m/z 252 in Figure 4.10b solves the problem.

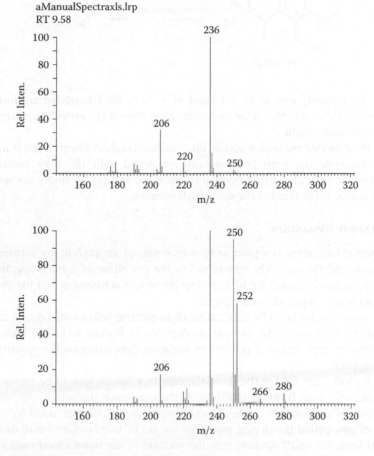

FIGURE 4.10 Mass spectra of O-methylpellotine ($M = 251$) produced by (a) electron bombardment and (b) chemical ionization.

The same two fragment ions at 236 and 250 are formed with equal abundance in the CI spectrum, reflecting loss of methane (CH_4) and hydrogen (H_2) molecules (Equation 4.5).

The low-intensity ions at M + 15 and M + 29 in the CI spectra are formed by addition of CH_3^+ and $C_2H_5^+$ to the molecule rather than of H^+, providing confirmation of the molecular weight.

CI can be carried out with a normal electron bombardment source that is modified to accommodate the high pumping rates required with the high reactant gas pressures. For normal electron bombardment work, the gas is simply not admitted, and the source is operated at the usual high vacuum.

ELECTROSPRAY IONIZATION

Electrospray ionization is a process by which ions of an analyte are formed in the liquid state and then quickly transferred to the gas phase of a mass spectrometer. This is particularly useful for high molecular weight substances and for analyzing the eluant from a liquid chromatograph.

The process is initiated by application of an electric field to the tip of a capillary containing a solution of the analyte, as depicted in Figure 4.11. The ions formed in this manner may be either positive or negative, depending on the polarity of the voltage applied.

The high electric field at the tip of the capillary causes ions to be formed in the liquid cone emerging from the capillary. At the same time, the nebulizing gas begins the process of evaporation of the solvent. Evaporation is further aided by a second stream of gas, called the drying gas. This stream of increasingly small droplets is directed toward a small opening into the vacuum of the mass spectrometer, where evaporation is completed, leaving individual ions that can be focused and separated according to their mass, producing the mass spectrum.

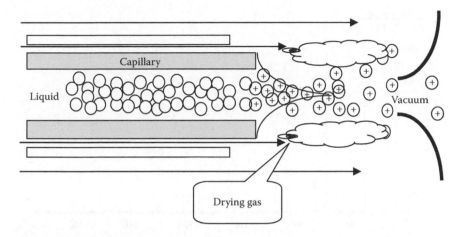

FIGURE 4.11 Schematic of ESI process.

Since ionization occurs in the liquid state, the specific ion formed is highly dependent on not only the analyte but also the solvent and other species present in the liquid. It must be remembered that the solvent will be ionized as well as any analyte that is present.

Positive ESI

In positive ion mode, M + H (M + 1) and M + Na (M + 23) ions may be formed from organic analytes. Since this is a very "soft" type of ionization, fragmentation is rarely seen, and the spectrum is dominated by these molecular ion species.

The mass spectrum in Figure 4.12 shows the positive spectrum background when a mixture of slightly acidified water and methanol is eluting from the capillary tip. The spectrum has many weak peaks, but the major ion masses are assigned in Table 4.1.

The following rules generally apply for positive ESI:

- M + sodium (M + 23) ions are favored from organic species containing multiple oxygen atoms.
- M + hydrogen (M + 1) ions are favored from basic amine species.
- Both ions will be formed from species containing both amine and oxygen functional groups.
- As the number of heteroatoms present in a molecule increases, the signal strength (or sensitivity) also increases.
- Since fragmentation is not present, isotope peak intensities take on added importance for identification purposes.

EXERCISE 4.2

Identify the compound giving the positive ESI spectrum in Figure 4.13.

- The peak at 193 Da (M + H or M + Na) indicates a molecular weight of either 192 or 170.

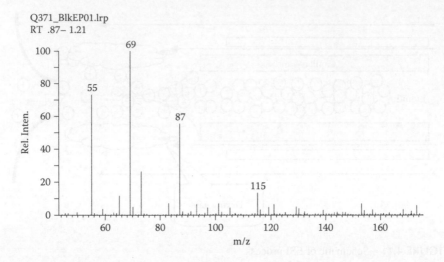

FIGURE 4.12 Positive ESI mass spectrum of acidified water and methanol (normal background spectrum).

TABLE 4.1
Background Ions from Positive ESI

Mass	Ion Species	Mass Calculation
55	$[CH_3OH + Na]^+$	$32 + 23 = 55$
69	$[(H_2O)_2 + CH_3OH + H]^+$	$2*18 + 32+1 = 69$
87	$[(CH_3OH)_2 + Na]^+$	$2*32 + 23 = 87$
115	$[(CH_3OH)_3 + H_2O + H]^+$	$3*32 + 18+1 = 115$

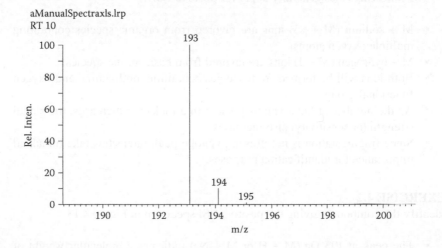

FIGURE 4.13 Exercise 4.2.

- The isotope peak at 194 indicates the presence of nine carbon atoms (10.3/1.1 = 9), which accounts for 108 Da (9*12 = 108).
- The even molecular weight indicates no nitrogen atoms or an even number of nitrogen atoms.
- If the molecular weight is 170, then the ion at 193 Da corresponds to M + Na.
  - That leaves 62 Da (193 − 23 − 108 = 62) to be accounted for.
  - $H_{30}O_2$ (30 + 2*16) is not possible since there are too many hydrogen atoms.
  - $H_{14}O_3$ (14 + 3*16) could account for this mass.
  - H_2O_4 (2 + 4*16) is not likely since there are so few hydrogen atoms.
  - $H_{18}ON_2$ (18 + 16 + 2*14) also could account for this mass but would likely give an M + H ion at 171 Da.
- If the molecular weight is 192, then 193 is M + H, indicating an amine compound with an even number of nitrogen atoms.
  - That leaves 84 Da (193 − 1 − 108 = 84) to be accounted for.
  - $H_{24}O_2N_2$ (24 + 2*16 + 2*14) is not possible since there are too many hydrogen atoms.
  - $H_{20}O_3N_2$ (20 + 3*16 + 2*14) is not likely since this composition would likely give an M + Na ion at 215 Da.
- **The most likely composition is $C_9H_{14}O_3$. This is the ESI spectrum of the same compound giving the EI spectrum in Figure 3.11.**

EXERCISE 4.3

Identify the compound giving the positive ESI spectra in Figure 4.14.

- The molecular weight is either 169 or 147.
- The isotope peak at 171 indicates the presence of 12 or 13 carbon atoms (13.6/1.1 = 12.4).

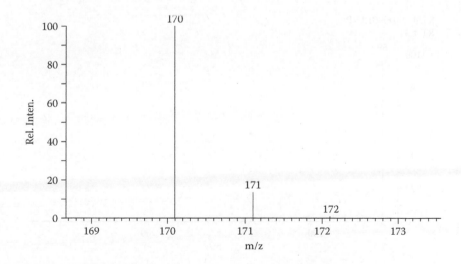

FIGURE 4.14 Exercise 4.3.

- Odd molecular weight indicates an odd number of nitrogen atoms.
- If the molecular weight is 147, then the 170 ion corresponds to M + Na.
  - Thirteen carbons are not possible since the mass is too high (13*12 = 156).
  - Twelve carbons are also not possible (12*12 = 144) since the remaining 3 Da could only be 3 hydrogen atoms.
- If the molecular weight is 169, then the 170 ion corresponds to M + H.
  - Thirteen carbons are not possible (13*12 = 156) since there is not enough mass left for a nitrogen atom.
  - Twelve carbons (12*12 = 144) leaves 25 Da to be accounted for.
  - $H_{11}N$ (11 + 14) accounts for this mass.
- **The most likely composition is $C_{12}H_{11}N$.** This is the ESI mass spectrum of diphenylamine.

Negative ESI

In negative ion mode, M – H (M – 1) is formed from acidic organic analytes. Since this is a very soft type of ionization, fragmentation is rarely seen, and the spectrum is dominated by these molecular ion species.

The mass spectrum in Figure 4.15 shows the background negative spectrum when a mixture of water and methanol, slightly acidified with acetic acid, is eluting from the capillary tip. The spectrum has many weak peaks but is dominated by the major M – H ion from acetic acid (see Table 4.2).

The following rules generally apply for negative ESI:

- M – hydrogen (M – 1) ions are favored from acidic organic species.
- Other negative ion species that are sometimes observed from nonacidic species are shown in Table 4.3.
- Since fragmentation is not present, isotope peak intensities take on more importance for identification purposes.

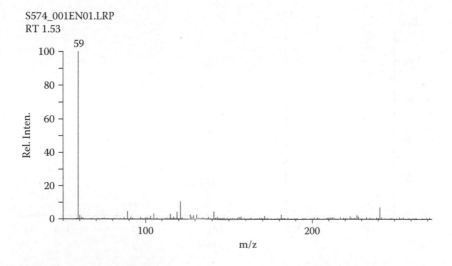

FIGURE 4.15 Negative ESI spectrum of water and methanol acidified with acetic acid.

TABLE 4.2
Background Ions from Negative ESI

Mass	Ion Species	Mass Calculation
59	[CH₃COO]⁻	$3 + 2*24 + 2*16 = 59$

TABLE 4.3
Negative Adduct Ions of Nonacidic Compounds

Adduct	Mass Observed
Cl⁻	$M + 35$
HCOO⁻	$M + 45$
CH₃COO⁻	$M + 59$
O₂⁻	$M + 32$

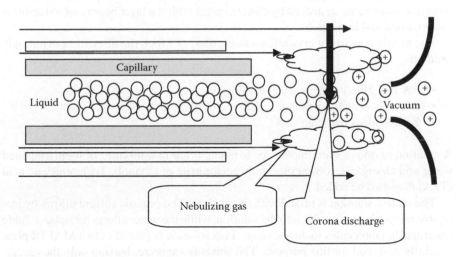

FIGURE 4.16   Schematic of APCI process.

ATMOSPHERIC PRESSURE CHEMICAL IONIZATION

APCI is a gas phase ionization process that creates ions at atmospheric pressure. A sample solution flows through a heated tube (>400°C), where it is volatilized and sprayed into a corona discharge with the aid of nitrogen nebulization. Ions are produced in the discharge and extracted into the mass spectrometer as depicted in Figure 4.16.

The corona discharge ionizes the solvent molecules in the tiny droplets as they pass by a tip held at very high voltage. In turn, the charge is transferred to any analyte molecules present in the solvent stream.

A positive APCI mass spectrum usually contains the quasi-molecular ion, $[M + H]^+$. However, APCI is a less-soft ionization technique compared to ESI and generates more fragment ions.

A negative APCI mass spectrum usually contains the $[M - H]^-$ ion. $[M + O_2]^-$ ions may also be observed.

This technique is used as an LCMS (liquid chromatographic-mass spectrometric) interface because it can accommodate very high (1 mL/min) liquid flow rates.

APCI is best suited to relatively polar, semivolatile samples.

MATRIX-ASSISTED LASER DESORPTION IONIZATION

MALDI is a soft ionization technique used for the analysis of biomolecules (biopolymers such as proteins, peptides, and sugars) and other large organic molecules, such as polymers, dendrimers, and other macromolecules, which tend to be fragile and fragment when ionized by more conventional ionization methods. It is most similar in character to ESI both in relative softness and the ions produced.

The ionization is triggered by a laser beam. A matrix is used to protect the macromolecule from being destroyed by direct contact with the laser beam and to facilitate vaporization and ionization.

The matrix consists of crystallized molecules, of which the three most commonly used are

- 3,5-dimethoxy-4-hydroxycinnamic acid
- α-cyano-4-hydroxycinnamic acid
- 2,5-dihydroxybenzoic acid

A solution of one of these molecules is made, often in a mixture of highly purified water and an organic solvent (normally acetonitrile or ethanol). Trifluoroacetic acid (TFA) may also be added.

The matrix solution is mixed with the analyte. The organic solvent allows hydrophobic molecules to dissolve into the solution, while the water allows for water-soluble (hydrophilic) molecules to do the same. This solution is placed onto a MALDI plate specially designed for this purpose. The solvents vaporize, leaving only the recrystallized matrix, but now with analyte molecules spread throughout the crystals. The matrix and the analyte are said to be cocrystallized in a MALDI spot.

The laser is fired at the crystals in the MALDI spot. The matrix absorbs the laser energy, becoming partially ionized. The ionized matrix then transfers part of its charge to the analyte molecules, thus ionizing them while still protecting them from the disruptive energy of the laser. Ions observed during this process consist of a neutral molecule [M] with an added or removed ion. Together, they form a quasi-molecular ion, for example, $[M + H]^+$ in the case of an added proton, $[M + Na]^+$ in the case of an added sodium ion, or $[M - H]^-$ in the case of a removed proton.

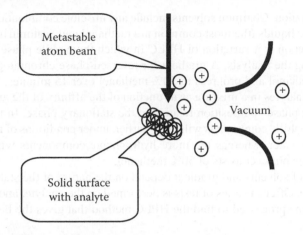

Metastable atom beam

Vacuum

Solid surface with analyte

FIGURE 4.17 Schematic of DART ionization process.

DIRECT ANALYSIS IN REAL TIME

DART is a technique used to ionize analytes on surfaces in open air. DART uses primary and secondary ionization brought about by metastable atoms of helium or molecules of nitrogen. The metastable atoms impinge on a surface containing analyte molecules, as depicted in Figure 4.17. The molecules are ionized in a complex process, generally producing ions similar to those of APCI. That is, in positive-ion mode, [M + H]$^+$ ions are formed, and in negative-ion mode, [M – H]$^-$ ions are formed.

When the DART ion source is placed near the opening to the mass spectrometer, the ions are sucked in by vacuum.

LIQUID CHROMATOGRAPHY-MASS SPECTROMETRY

INTRODUCTION

High-performance (or high-pressure) liquid chromatography (HPLC) is a form of column chromatography used to separate, identify, and quantify compounds. HPLC utilizes a column that holds chromatographic packing material (stationary phase), a pump that moves the mobile phase(s) through the column, and a detector that shows the retention times of the analyte molecules. An HPLC system is used to introduce samples to a mass spectrometer and forms the combined technique known as liquid chromactrography - mass spectrometry (LCMS).

The sample to be analyzed is introduced into a stream of mobile phase. The analyte molecules are swept through the column by the mobile phase and retarded by specific chemical or physical interactions with the stationary phase. The amount of retardation depends on the nature of the analyte, stationary phase, and mobile phase composition. The time for a specific analyte to traverse the length of the column is called the retention time. Various stationary phase materials result in predictable interactions with specific chemical families, thus giving the desired degree of chemical separation. Common solvents include any miscible combination of water or various organic liquids (the most common are methanol and acetonitrile).

Gradient elution is a variation of HPLC in which the mobile phase composition is varied during the analysis. A gradient for reversed-phase chromatography might start at 5% methanol and progress to 50% methanol over 15 minutes. The gradient separates the analytes in a mixture as a function of the affinity of the analyte for the current mobile phase composition relative to the stationary phase. In this example, the more hydrophilic compounds will elute earlier, under conditions of relatively low methanol/high water, whereas the more hydrophobic components will elute later, when the mobile phase consists of 50% methanol.

The choice of solvents and gradient depend on the nature of the stationary phase and the analyte. Often, a series of tests is performed on the analyte, and a number of trial runs may be processed to find the HPLC method that gives the best separation of components.

The combination of HPLC and mass spectrometry was impossible for many years due to the difficulty in removing the solvent prior to the analyte entering the high vacuum of the mass spectrometer. However, with the advent of the atmospheric pressure ionization techniques discussed previously, the two instruments have been combined to give the relatively new technique of LCMS. The eluant from the HPLC is passed into one of the atmospheric pressure ion sources, where ions are produced and most of the solvent is rapidly evaporated. The ion cloud then passes through a small aperture into the vacuum of the mass spectrometer, where the solvent is totally removed, and the ions can be focused and mass analyzed in the normal manner.

Some limitations to the normal array of HPLC methodologies exist for LCMS applications:

- The mobile phase flow rate generally must be kept below 1 μL per minute.
- Buffers must be limited to those reagents that are easily evaporated.
  - Acceptable
    - Ammonium acetate
    - Ammonium formate
  - Not acceptable
    - Phosphates
    - Other inorganic buffers
- The solvents must be ionizable by the ionization method being used.
  - Acceptable
    - Water
    - Methanol
    - Ethanol
    - Isopropanol
    - Tetrahydrofuron (THF)
    - Acetonitrile
  - Not acceptable
    - Hydrocarbons
    - Chlorinated solvents

APPLICATIONS

Applications for LCMS are diverse, as evidenced by this partial list:

- Neuropeptides
- Combinatorial chemistry
- Environmental chemistry
- Biochemical and biotechnological applications
- Rapid drug discovery
- Drug design
- Food science and technology
- Drug and cosmetic excipients
- Forensic applications
- Drug screening and analysis
- Peptides and proteins
- Biological macromolecules
- Agricultural chemistry
- Biology and medicine
- Cancer research
- Neuropeptide research
- Polymers
- Enzymology
- Vitamins

When the analytes to be identified or quantified are polar or have a high molecular weight, LCMS should be considered over other mass spectrometric techniques. Some examples follow.

Example 4.1

The pharmaceutical A with the composition $C_{20}H_{19}ClFNO_4$ produces an unknown metabolite B (Equation 4.5). Identify the metabolite.

$$A \xrightarrow{\text{metabolism}} B \tag{4.5}$$

The positive ESI mass spectrum of A shown in Figure 4.18 shows the expected M + Na ion at 414 Da (Equation 4.6) and the chlorine isotope peak at 416 Da.

$$20*12 + 19*1.007825 + 34.9689 + 18.9984 + 14.0031$$
$$+ 4*15.9949 + 22.9898 = .0885 \tag{4.6}$$

The chromatogram obtained for metabolite B by LCMS using positive ESI is shown in Figure 4.19. Note that the high background (see Figure 4.12) of the TIC makes it difficult to detect chromatographic peaks. A reconstructed chromatographic type known as the component detection algorithm (CODA) extracts chromatographic peaks from LCMS data. This technique creates RICs for every mass

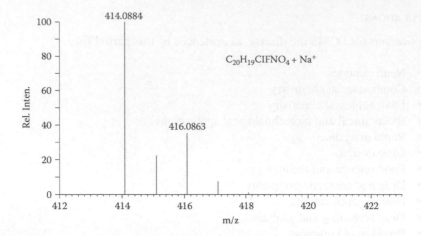

FIGURE 4.18   Positive ESI mass spectrum of compound A.

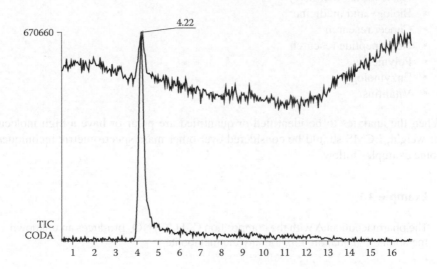

FIGURE 4.19   TIC and CODA chromatograms for metabolite B.

and then applies statistical analyses to the chromatograms to calculate a quality score for each individual chromatogram. Those chromatograms that give high scores, or actually contain chromatographic peaks, are summed to produce the CODA chromatogram. The mass spectrum of metabolite B is shown in Figure 4.20.

The accurately measured mass for the M + Na ion is 416.1043, corresponding to two more hydrogen atoms than present in A (Equation 4.7).

$$416.1043 - 414.0885 = 2.0158 \sim 2*1.007825 \qquad (4.7)$$

Metabolite B contains two more hydrogen atoms than compound A and is likely to be an alcohol derivative of a keto function that is present in A.

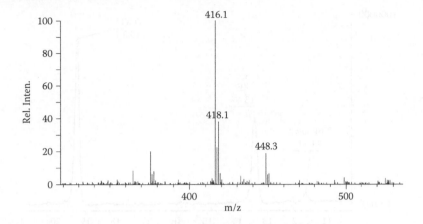

FIGURE 4.20 Positive ESI mass spectrum of metabolite B.

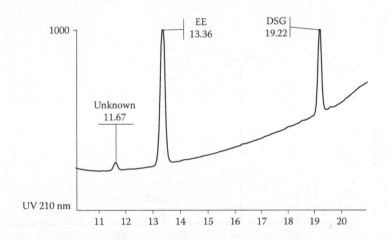

FIGURE 4.21 UV chromatogram from liquid chromatographic analysis of pharmaceutical.

EXERCISE 4.4

An impurity was detected during routine liquid chromatographic screening using UV detection in a pharmaceutical containing two active ingredients (see Figure 4.21).

The following data related to this impurity were obtained:

- No signal by negative ESI or negative APCI. This indicates the absence of acidic or phenolic functional groups.
- No signal by positive ESI, indicating the absence of polar functional groups.
- Positive APCI detected a signal eluting at the expected retention time (see Figure 4.22).
- The positive APCI mass spectrum (Figure 4.23) contains an M + H ion at 195 Da, accurately measured to be 195.1021 Da.

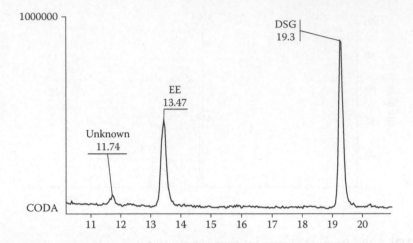

FIGURE 4.22   CODA chromatogram from LCMS analysis of pharmaceutical using positive APCI.

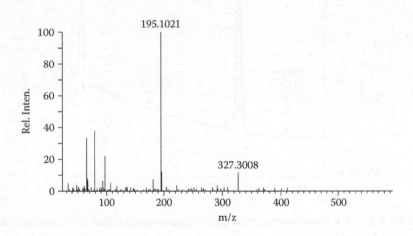

FIGURE 4.23   Positive APCI spectrum of impurity.

- The elemental composition of the impurity is shown to be $C_{11}H_{14}O_3$ (Equation 4.8).

$$11*12 + 15*1.007825 + 3*15.9949 = 195.10208 \qquad (4.8)$$

- The impurity was isolated and its EI spectrum measured (Figure 4.24).

PROBLEM 4.2
Identify the impurity.

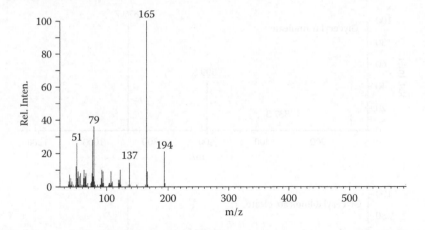

FIGURE 4.24 Positive EI mass spectrum of impurity.

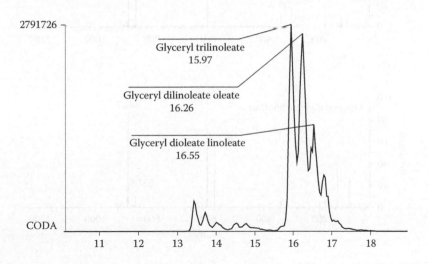

FIGURE 4.25 Positive APCI chromatogram of oil stain.

EXERCISE 4.5

An oil stain related to a criminal investigation was analyzed by LCMS using positive APCI (see Figure 4.25).

The molecular weights of the major components (see Figure 4.26) correspond to the triglycerides glyceryl trilinoleate, glyceryl dilinoleate oleate, and glyceryl dioleate linoleate. The stain is therefore identified as a vegetable oil. These results are then compared to the analytical results from several known vegetable oils (Figures 4.27 and 4.28). Figure 4.25 most closely resembles Figure 4.27a, and the stain is identified as a corn oil.

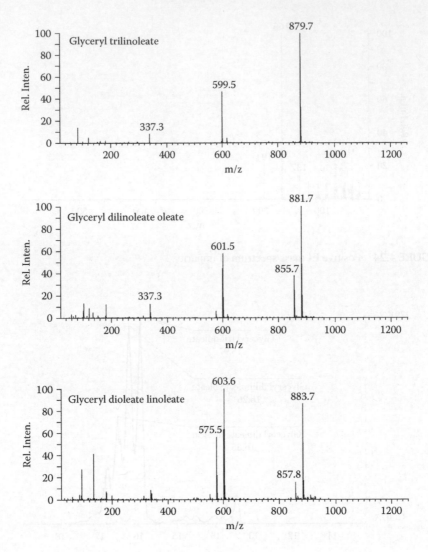

FIGURE 4.26 Positive APCI mass spectra of triglycerides.

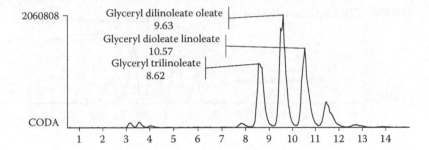

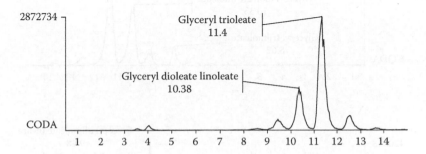

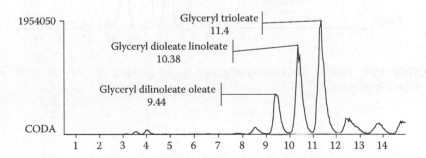

FIGURE 4.27 Positive APCI chromatograms for (a) corn oil, (b) peanut oil, and (c) olive oil.

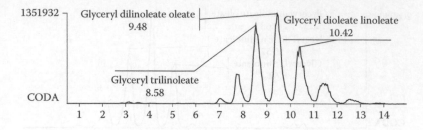

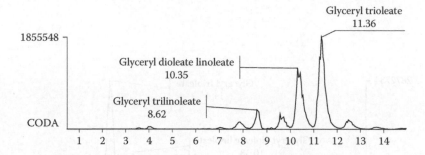

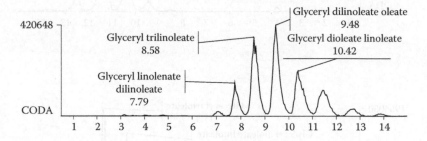

FIGURE 4.28 Positive APCI chromatograms for (a) soybean oil, (b) canola oil, and (c) vegetable shortening.

Appendix I Answers to Problems

CHAPTER 2

2.1	Fig. 2.4	acetone
	Fig. 2.5	benzaldehyde
	Fig. 2.6	dimethyl sulfoxide
	Fig. 2.7	terphenyl
	Fig. 2.8	bromocyclohexane
	Fig. 2.9	chloroform
2.2	Fig. 2.16	acetoin
2.3	Fig. 2.17	acetone oxime
2.4	Fig. 2.24	4-methyl-4-heptanol
2.5	Fig. 2.35	4-methyl-2-pentanone
2.6	Fig. 2.36	2-methyl-3-pentanone
2.7	Fig. 2.38	4-isobutylquinoline
2.8	Fig. 2.51	dichloromaleic anhydride
2.9	Fig. 2.52	cyclopropyl-methyl ketone
2.10	Fig. 2.53	3-hexanone
2.11	Fig. 2.57	biphenylene
2.12		SO_2
		CF_3
		C_3H_6Br or C_2H_2OBr
		$C_3H_5Cl_2$ or C_2HOCl_2
2.13	Fig. 2.61	2-methyl-2-heptanol
	Fig. 2.62	3-ethyl-3-hexanol
	Fig. 2.63	3-methyl-3-hexanol
2.14	Fig. 2.69	acetylsalicylic acid
	Fig. 2.70	thiosalicylic acid
2.15	Fig. 2.72	n-butyl (o-methylamino) benzoate
2.16	Fig. 2.73	4-methylbenzyl 4-methylbenzoate
2.17	Fig. 2.78d	4-phenyl-1-butanol
2.18	Fig. 2.79	3-methyl-2-butanone
2.19	Fig. 2.80	3-methylbutyraldehyde
2.20	Fig. 2.81	furfural
2.21	Fig. 2.82	p-bromophenol
2.22	Fig. 2.83	methyl 3-hydroxybenzoate
2.23	Fig. 2.84	divinyl ether
2.24	Fig. 2.85	benzil
2.25	Fig. 2.86	fluoroacetone
2.26	Fig. 2.87	6-methylsalicylic acid

2.27	Fig. 2.88	chloroacetic acid
2.28	Fig. 2.89	3, 5-dinitrotoluene
2.29	Fig. 2.90	2-methoxyethanol
2.30	Fig. 2.91	4-heptanone
2.31	Fig. 2.92	N-ethyl trifluoroacetamide
2.32	Fig. 2.93	ethyl cinnamate
2.33	Fig. 2.94	fumaric acid
2.34	Fig. 2.95	trifluoroacetone
2.35	Fig. 2.96a	2-tert-butyl-5-methylanisole
	Fig. 2.96b	ortho-n-pentylanisole
	Fig. 2.96c	para-n-pentylanisole
2.36	Fig. 2.97	3-(4-aminobutyl)-indole
2.37	Fig. 2.98	trichloroacetic acid
2.38	Fig. 2.99	3-methylquinoline
	Fig. 2.100	7-methylquinoline
	Fig. 2.101	2, 4-dimethylquinoline
	Fig. 2.102	7-ethylquinoline
2.39	Fig. 2.103	dimethyladipate
2.40	Fig. 2.104	n-butoxy-trimethylsilane
2.41	Fig. 2.105	hexanoic acid
2.42	Fig. 2.106	1, 2, 3-trimethoxypropane
2.43	Fig. 2.107	p-bromoacetanilide
2.44	Fig. 2.108	p-chlorophenoxyacetic acid
2.45	Fig. 2.109	4-undecanone
2.46	Fig. 2.110	o-aminoanisole
2.47	Fig. 2.111	3-heptanone
2.48	Fig. 2.112	dibenzothiophene oxide
2.49	Fig. 2.113	dibenzoylamine
2.50	Fig. 2.114	chlorpropham
2.51	Fig. 2.115	iodoacetamide

CHAPTER 3

3.1	1	$C_8H_8O_2$ (methyl benzoate)
	2	$C_9H_8N_2$ (4-aminoquinoline)
	3	$C_4H_8O_3$ (methyl 2-hydroxypropionate)
	4	$C_{10}H_7Br$ (bromonaphthalene)
3.2		59.0452 (^{13}C isotope) 60.0461 (^{17}O isotope)
3.3	Fig. 3.8	2-butylfuran
3.4	Fig. 3.9	2,3-dimethyl-2, 3-butanediol
3.5	Fig. 3.10	2-ethylpiperidine
3.6	Fig. 3.11	2-(2-methyl-1,2-dihydroxybutyl) furan

CHAPTER 4

4.1	Fig. 4.3	ethylbenzene
4.2	Fig. 4.24	3,4-dimethoxypropiophenone

Index

T - #0141 - 101024 - C0 - 234/156/10 [12] - CB - 9781466595842 - Gloss Lamination